REVISED FIRST EDITION

DISSECTION SIMPLIFIED

A LAB MANUAL

FOR INDEPENDENT WORK IN HUMAN ANATOMY

BY DR. DANIELLE DODENHOFF
CALIFORNIA STATE UNIVERSITY—BAKERSFIELD

Bassim Hamadeh, CEO and Publisher
Kassie Graves, Director of Acquisitions
Jamie Giganti, Senior Managing Editor
Jess Estrella, Senior Graphic Designer
Seidy Cruz, Senior Field Acquisitions Editor
Natalie Lakosil, Licensing Manager
Allie Kiekhofer and Kaela Martin, Associate Editors
Mandy Licata, Interior Designer

Cover image copyright© iStockPhoto / dimid_86

Printed in the United States of America

ISBN: 978-1-5165-0897-6 (pbk) / 978-1-5165-0898-3 (br)

cognella® | ACADEMIC PUBLISHING

CONTENTS

ACKNOWLEDGMENTS

As with any textbook, more colleagues helped than can be acknowledged during the development of this lab manual. However, I want to acknowledge several individuals that significantly contributed to its development over the last 17 years. I began teaching anatomy as a graduate student because there were few students that wanted to invest the considerable time needed to learn and teach the material. I found once I had learned the material, the challenge was to help the students do the same in less time (I, of course, took several terms to master the course material). I found the biggest challenge was to have the students work with the material outside of the scheduled class time. Since most specimens could not leave the lab, students were unable or reluctant to come into the lab for the necessary study time. With my colleague, Dr. Robert Stark, we began developing an interactive website the students could use to review the material when they were not in lab. This full year endeavor is the source of most of the images for the dissections. To develop the images and handouts we required equipment and lab space. Therefore, I wish to thank the department of EEOB at The Ohio State University for use of their facilities. After developing the website, I began to notice students did not know how much effort was necessary to memorize the course material. Most students were surprised that although they knew the structures while studying, they were unable to recall the information during exams. Apparently the worksheets in the lab manual did not help the students with practical examinations. Therefore, I began developing worksheets to help with lab material and the more difficult lecture topics. I also noticed that my students had difficulty with formal dissection manuals and were unsuccessfully relying on dissection demonstrations at the beginning of lab. To address this problem I typed up step-by-step instructions and "helpful hints" to summarize each lab. These worksheets and instructions have been expanded and developed while teaching Human Anatomy at California State University, Bakersfield. I wish to thank the faculty of the Biology department at CSUB for their assistance with supplies and specimens. I also wish to thank the students for their invaluable feedback over the years. The term lists of essential anatomy material were refined through collaboration with professors Dr. Ken Gobalet and Dr. Todd McBride. I have been fortunate to have supportive colleagues during the development the Human Anatomy course (Bio 250). I hope this manual significantly aids students in mastering Human Anatomy allowing them to pursue their academic goals.

AN INTRODUCTION TO HUMAN ANATOMY

This lab manual is the result of many years of teaching anatomy labs and watching the students try to complete dissections without using a lab manual. Students did this primarily because the traditional manuals did not describe in common language how to find structures. As I began developing worksheets to help students through the dissections, I realized that the traditional lab manual format had to change. Most manuals are in the same format. This manual is meant to be used with a traditional textbook but is designed to help students independently investigate anatomy. Recall is more efficient when associated with previously learned material or otherwise familiar facts about an anatomical structure. This learning principle is the reason some students seem to learn the material more easily than others. The variety of scientific background for students taking anatomy was also a motivation for this manual. Since many students will be unfamiliar with scientific terminology the following worksheets contain suggestions for how to better study the materials for efficient recall.

Anatomy is a subject I refer to as a "necessary evil"; to learn anatomy is very time consuming. Without knowing anatomical structures advanced studies in the field of health becomes difficult. Few students begin their educational career with the final goal as an anatomist. Most of the students taking anatomy are doing so because it is required to continue with advanced studies in a health related career. Anatomy courses vary in difficulty depending on the college, instructor, and major of students required to take the course. It is understandable that Gross Human Anatomy for aspiring physicians would be difficult, but most students taking an introductory level anatomy course are not in medical school. So many of you might be questioning the validity of learning a seemingly impossible number of terms and structures you will never remember or need to know once you graduate. I can explain the necessity for learning detailed information for health professionals (and yes even some health related business degrees are requiring anatomy). Understanding and practice of health procedures requires experience with the composition of the body. For example, when giving a shot or drawing blood a person should be aware of the possible tissues that will be damaged and what is underlying the skin. But let us say you are going to be an athletic coach and are not allowed to treat student injuries. You will still need to be aware of the human body and its limitations. If there is an injury knowing more about anatomy would give you an appreciation for adhering to school policies about dealing with injuries, rather than ignoring an injury or diagnosing the problem (something you should never do without a medical degree). To gain an appreciation for the composition of the human body, we will investigate each body system and its components. This manual is an introduction to anatomy so you will not have to learn all structures of the human body (although you might think you are at times). The lists included with each worksheet section were complied to contain a learnable amount of structures within a week and half time and give an accurate mental image of each system. In true education tradition, you can expect to remember approximately 10–15% of the material you are required to learn during the term. This minor amount of retained material can only be accomplished by requiring more detail during the term. So even if you "brain dump" the material after each exam, by the end of the term you will appreciate more the complexity of fixing damaged tissues and organs. For example, why does it take longer for a ligament to heal than a fractured bone?

As we all have the goal of achieving academic success in anatomy, I understand your concern will be to maximize your grade in a course. This is not only ok, but expected. It is the professor's job to make sure the assessments given during a course accurately reflect your understanding of anatomical material. Your job will be to read the study hints in the manual to help with learning the information for each exam. This course is different from most courses you will take at an educational institution. Most courses contain approximately 25–40% new material, which is why you may take a course and not have to study much before an exam and still do well on the exam. This is standard for education because students learn by repeating lessons in each successive class. The difficulty with anatomy is that most of the information has never been covered in previous classes. Therefore, you will most likely encounter approximately 80–95% new information. If you have recently taken a general biology class you may only have about 70% new information. The last time you were asked to learn this much new information this quickly was when you first entered school. However, young children are also not tested on all of the material in two weeks with consequences to their educational success. The course material is not difficult to understand, but it is time consuming to learn. "Cramming" terms two days before an exam is not an effective method for successful recall. The technique of cramming usually results in knowing the terms but improperly applying them during practical examinations. The other drawback is the number of questions that can be missed. It is a numbers game. To achieve 70% you have to get 70% of the points. I know this seems obvious, but I point it out because 50% of anatomy information will be a lot of information. In other courses this amount of information would be sufficient. However, many students get 5 or 6 questions correct on a 10 pt quiz and think they know the material well enough for an exam (but it is not 70%). I have included this warning about anatomy, not to discourage, but to forewarn. I hope the knowledge in this manual will be enjoyable and applicable. After all we do not come with an instruction manual so it should be interesting to learn more about how your body structured. Fair warning it will be a lot of work for the entire term from the first to the very last day. But I hope this manual will guide you to successfully completing this material and to gain appreciation for the human body once your course is complete.

WORKSHEET I

Skeletal System—Axial

The bones and structures listed on the first page are required for the laboratory portion of the skeletal system. The exam covering this material will require the identification of articulated and disarticulated bones. Many of the skull bones will be articulated; however, the disarticulated specimens available in the lab will also be on the exam. To find each individual structure, you will have to read the chapter and find the structures using the textbook or lab manual supplement [the time and effort taken to look up and read the description for each structure will aid in learning the structures for the lab practical].

AXIAL SKELETON

Skull Bones and Structures:

Cranial bones:	Bone structures	Facial bones:	Bone structures
Frontal		Palatine	hard palate [maxilla also contributes]
Parietal		Maxilla	hard palate molars, premolars, canine and incisors
		Vomer	
Occipital	external occipital protuberance occipital condyles	Zygomatic	zygomatic arch [both the zygomatic and temporal contribute a section of the arch]
Temporal	zygomatic arch mastoid process external and internal auditory meatus mandibular fossa petrous	Mandible	condylar and coronoid processes molars, premolars, canine and incisors
Sphenoid	sella turcica [for the pituitary gland]	Lacrimal Nasal	
Ethmoid	nasal conchae [superior and middle] cribriform plate [with olfactory foramina] perpendicular plate	Inferior nasal conchae [not visible on many of the skulls]	

Sinuses:	Foramen (or canal)		
frontal sphenoid ethmoid maxilla	foramen magnum olfactory foramina foramen rotundum hypoglossal foramen optic foramen	jugular foramen carotid foramen(canal) superior orbital fissure stylomastoid foramen foramen ovale mandibular foramen	

Axial Bones and Structures:

Hyoid	Fetal skull
Auditory ossicles malleus incus stapes	Identify the major cranial bones on the fetal skull Fontanels: frontal, sphenoidal, and mastoid

Vertebrae cervical (7)—atlas, axis, and 5 unspecialized thoracic (12)—with articular facets for ribs lumbar (5) sacral (5 fused) - sacroiliac joint coccyx	Vertebral features body [centrum] spinous process vertebral foramen transverse process superior articular facet inferior articular facet
Sternum manubrium body xiphoid process	Ribs Costal cartilages

This worksheet is designed to point out methods of learning the skeletal system. Therefore, many of the questions, although not specifically exam questions, are intended to guide you while learning the bones and structures. Although you will spend much of your lab time looking up the structures in the textbooks, to remember the structures for the lab exam, it is also helpful to learn how the structures are associated to one another. The additional information about how the structures are associated and function will aid in your ability to recall the correct term during the fill-in-the-blank section of the exams. First before answering the worksheet questions, use the textbook to find each individual structure and **read the descriptions of the terms** in each chapter [the time and effort taken to look up and read the description for each structure will aid in learning the structures for the lab practical]. The more associated memory a term has the easier it is to recalled from long term memory. Categorizing information adds to the associated memory of a term. For example which is easier to remember 661654225 or (661) 654-2225? Because you need to categorize the terms it would be helpful if you use the categorization method that will help with recognition on a practical lab exam. Therefore the following paragraph lists terms you need to know to categorize the bone structures and page 7 of this exercise suggests a study hint to increase your ability to correctly categorize the skeletal terms.

The study of anatomy requires the use of standardized terms for regions of the body and locations of anatomical structures relative to one another. These terms are explained in the textbook, usually in the first chapter. You should know what the following terms mean: medial, lateral, sagittal, mid-sagittal, inferior, superior, anterior, posterior, proximal, and distal. You should also know the proper description for anatomical position and why it is used for anatomical reference. Although exams will not include definitional questions, these anatomical terms will be used in questions. Therefore, you will need to familiarize yourself with these terms. There are questions embedded throughout the skeletal and tissue worksheets to aid in learning the anatomical terms. Please note that this worksheet is not an exhaustive list of questions and will not include questions about all of the listed structures. These questions are designed to ask about common misconceptions and missed terms; if there is not a question on a structure, you will still have to be able to identify the structure on a practical lab exam.

Axial Skeleton Questions

[use the specimens in the lab and the textbook to answer the following questions. If you are unfamiliar with a term or structure please use the index provided with the textbook].

1. Which of these bones belongs to the cranium, not to the face? [nasal, palatine, sphenoid, zygomatic, and mandible]

2. Which of the 6 cranial bones that constitutes the external surface of the skull is not classified as a flat bone? _____
    Refer to illustrations; what color indicates this bone in all four figures? _____
    [note this bone composes most of the cranial "floor" of the skull]. This question is used to point out that most textbooks will fill in each bone with the same color in all illustrations. This allows the reader to identify different areas on the same bone, rather than a single labeled point.

3. The only movable bone that is part of the adult human skull is the _____.

4. The mandibular fossa is located in which bone? _____. This fossa [shallow depression] is the surface where the _____ process of the mandible articulates with the skull.

5. The bone that forms the shape of your face (wide or narrow) is your "cheekbone" or the _____ bone.

6. If you fall over backwards and hit the ground with the back of your head, you will "see stars." The skull bone that hit the ground is the _____ bone.

7. The inner and middle ear structures are housed within the _____ portion of the temporal bone. Also notice that the internal auditory meatus is located in the middle of this small bone of the temporal and is a good landmark for discerning the internal auditory meatus from the hypoglossal canal or foramen ovale [hint: this bone is often covered in the auditory section of a textbook].

8. The cranial bone that contains the foramen magnum is the? _____

9. The bony orbit (eyesocket) is made up of parts of seven different bones. Which of these facial and cranial bones do not contribute to orbital structure; nasal, lacrimal, sphenoid, zygomatic, or maxilla?

10. If the hard palate does not fuse during development, the result is termed "cleft palate." Therefore a cleft palate results when the _____ and _____ fail to fuse in the middle.

11. True or false, known as the "soft spot," the anterior or frontal fontanel is the largest of the six fontanels.

12. The fontanels are soft because they are not composed of bone but are instead composed of _____ _____ tissue.

13. The triangular-shaped bone that contributes to the ventral portion of the nasal septum is the _____.

AXIAL SKELETON

Vertebrae Questions:

14. Which cervical vertebra is called the "axis" because it allows the head to rotate: C1 or C2 or C3 or C5 or C9?

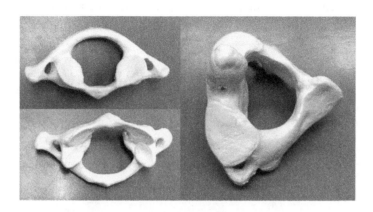

15. The bony ridges or projections you feel when you run your fingers along your spine are actually what structures? _____

16. Which type of vertebra have transverse foramen?

17. The demifacets are located on which vertebrae? These vertebrae also have greatly reduced lateral processes and long prominent spinous processes.

18. Also notice which structures articulate at the demifacts. The demifacts are located on which portion of the vertebrae? _____ and articulate with which structures?

19. Which vertebrae have a large central body with smaller transverse processes?

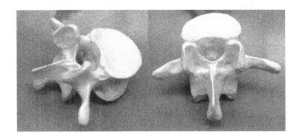

20. Which vertebrae are fused together and share an articulation with the iliac crest of the ilium?

21. Which fused set of vertebrae have reduced or lost a spine, vertebral arch, and transverse process?

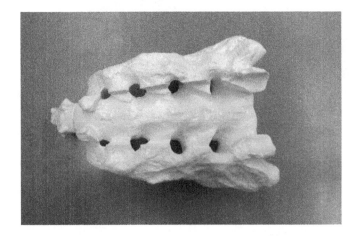

The following questions are designed to test your understanding of anatomical directions.

22. The sacral vertebrae are located superior or inferior to the thoracic vertebrae?

23. The humerus is located medially or laterally to the thoracic vertebrae?

Use the following questions to identify some important articulations.

24. What two projections of the skull articulate with the atlas (cervical) vertebrae?

25. A rib articulates with what bone structures on the thoracic vertebrae?

26. The only bone in the body that does not articulate with any other bone is the _____ bone.

27. The occipital condyles articulate with which cervical vertebrae? _____

Foramen Questions:

28. Which foramen can be seen in an inferior view of the ventral skull?

28a. Which foramen has the dowel exiting in view of the foramen magnum?

28b. Which foramen does not have a dowel exiting in view of the foramen magnum?

29. If you insert a wooden dowel through the optic foramen [from the internal superior skull view] the dowel will exit through which structure? _____.

30. Which foramen can be seen only in a superior internal skull view?

30a. The dowel marked Figure 1.2 D is most likely in the _____ although the actual foramen is not in view.

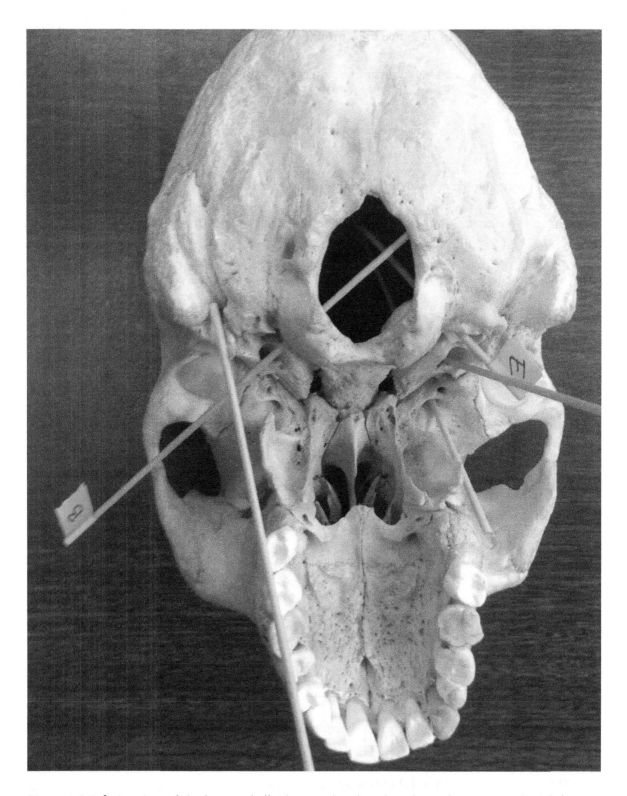

Figure 1.1 Inferior view of the human skull. The wooden dowels indicate foramen on the lab list.

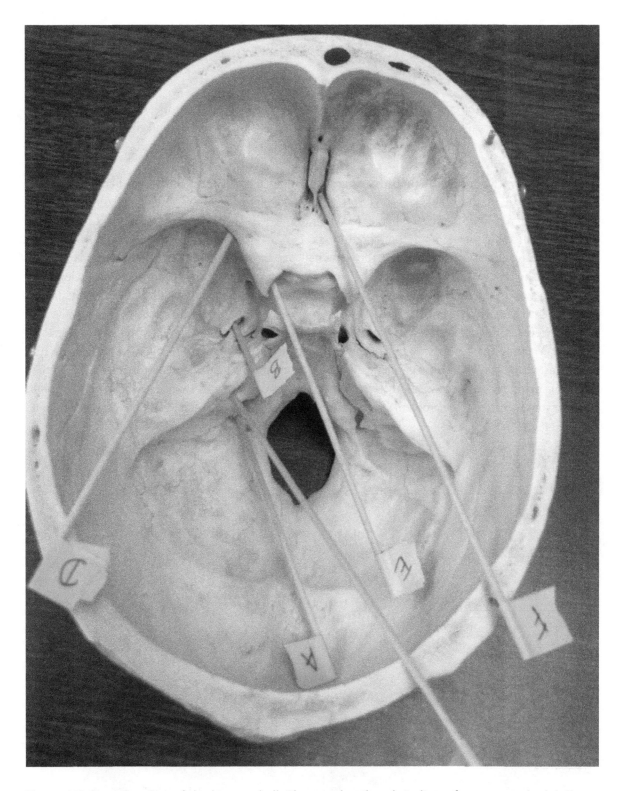

Figure 1.2 Superior view of the human skull. The wooden dowels indicate foramen on the lab list.

Study Tip: The practical portions of the exams are fill-in-the-blank questions; therefore studying by using matching and recognition of terms is not sufficient. A single practical question will require **3 steps**:

1. Identification of the specimen and which view is presented [lateral, medial, ventral, or dorsal]
2. Recalling a short list of possible structures found in the view presented, and finally
3. Filling in the term for the structure [spelled correctly].

The most efficient study method that increases memory recall for a fill-in-the-blank practical is repetition of **fill-in-the-blank views**. <u>If you study by recognition or by verbal quizzing you might have difficulty recalling the term when asked the same structure under test conditions</u>. I recommend studying using daily [or every other day] quizzing.

1. Create blanks of each skull view: The first challenge in learning anatomical structures is learning the terms. The best method for learning the terms is to organize them into their anatomical views. For example, a way to organize the structures by view would be to use blank skull pictures of the lateral, medial, sagittal, inferior, superior, and anterior views and list to the side all the structures that can be seen in these views. 2. White out any identification marks and draw blank lines to all structures required for the view. Make at least 6 photocopies of each view. Each study day try to fill in all of the structures. After a few study days, you will have a good idea of which structures you are not remembering. After 3 study days you should target the consistently missed structures by developing a story or association that might increase your memory of the problem structures. This may seem like too much work but each view should

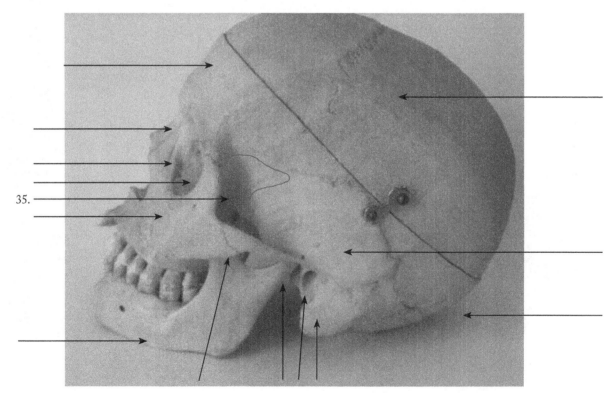

Figure 1.3 Lateral Skull View—This picture is a sample of a lateral view.

take only 5 to 10 minutes to fill out each study day [and you do not have to do each view every day]. Additionally this method of study is more effective than a 4-hour study session the night before since it requires you to recall the information several times [rehearsal of a memory path increases your ability to recall that memory]. Many students make the mistake of studying for 2 hours and thinking they have learned the structures. I hear from students all the time, "but I studied for hours". For recall memory it is not the amount of time with the material, it is the number of different times you require yourself to recall the information. However, if information is recalled within a 2-hour study session, the brain "considers" this as one time (even if you ask yourself the term 10 times during the 2 hours). You will have studied less than a student who requires recall each day for 10 days. (Each daily recall is only about 1 minute of study, but you will learn the structure better than if you study 3 times for 2 hours each.)

The structures are difficult to see because this is a picture of a real skull not an outlined illustration from the textbook. Using difficult pictures will force you to identify structures by their location relative to other structures rather than by the shape or blue dot you learn on a particular specimen. For example, line 35 is pointing to a region of the skull that you should be able to narrow down to a few possibilities because there are only a few structures in this area; the temporal, the sphenoid, the parietal, or the zygomatic. The ethmoid is eliminated from the list because it is not visible in a superficial lateral view. After determining the possibilities then you will have to learn characteristics of each that will narrow

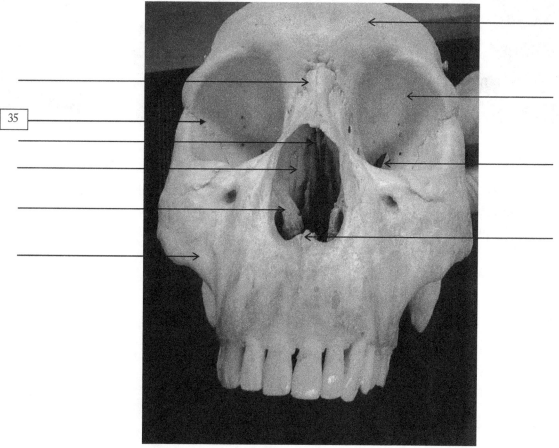

Figure 1.4 Frontal skull view of the human skill. Identify the individual parts of the ethmoid and other bones in view.

the list to one correct answer. Such as the temporal bone does not include the side depression around the orbit and the parietal bone ends before the temporal bone in the lateral view and the zygomatic bone is part of the cheek area not the lateral main skull. All of these characteristics leave the sphenoid as the only possible bone. In addition you need to learn shapes created by suture lines. Notice the light blue near the sphenoid outlines the sutures between the sphenoid, frontal, and parietal bones. These suture lines show the boundaries of the bones. You have to look at different views of the skull because bones like the sphenoid can be tagged in different views and will not appear to be the same bone if you do not know the boundaries of the bone. For example, find the sphenoid in the frontal view (Figure 1.4). Notice it looks different than the sagittal view of the sphenoid bone.

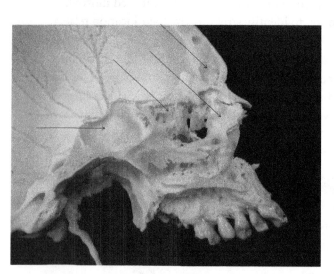 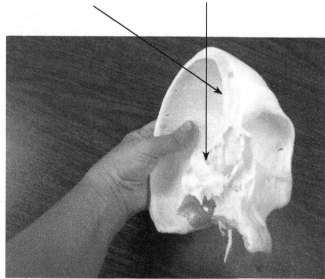

Figure 1.5 Mid-Sagittal view of skull.

Locate the internal nasal conchae (superior and middle that are part of the ethmoid bone). Identify the multiple cavities of the ethmoid sinus. The sphenoid sinus is located inferior to the sella turcica, which distinguishes it from the ethmoid sinus. The perpendicular plate is missing from this specimen (most specimens in the lab do not have the perpendicular plate).

Figure 1.6 Auditory ossicles, malleus, incus and stapes.

These small bones are housed in the petrosal portion of the temporal bone. Find the petrosal bone on the superior skull view. The petrosal has the internal auditory meatus and is a good method of identifying that foramen correctly.

WORKSHEET II

Skeletal System—Appendicular

The bones and structures listed on this page are required for the laboratory portion of the skeletal system. The exam covering this material will require the identification of articulated and disarticulated bones. An * indicates the bones you should be able to identify as being from the right or left side of the body.

APPENDICULAR SKELETON

Upper arm and Pectoral (shoulder) girdle		Lower limb and pelvic (hip) girdle	
clavicle		*ilium	crest, articular (auricular) surface, acetabulum, anterior superior iliac spine, anterior inferior iliac spine,
*scapula	glenoid cavity, spine, coracoid process, acromion process supraspinous fossa, infraspinous fossa, subscapular fossa	pubis	pubic symphysis
*humerus	head, greater and lesser tubercles, medial and lateral epicondyles olecranon fossa	ischium	ischial tuberosity, obturator foramen
ulna	olecranon process, trochlear notch radial notch, styloid process	*femur	head, neck, greater and lesser trochanters, medial and lateral condyles medial and lateral epicondyles, patellar surface
radius carpals	head, radial tuberosity, styloid process	patella *tibia	tuberosity, medial malleolus, lateral and medial condyles
metacarpals phalanges	proximal, middle, and distal	fibula tarsals metatarsals phalanges	head, lateral malleolus talus, calcaneus

Activities:

1. Identify all listed bones and structures.
2. Identify articulations between Appendicular structures.
3. Identify characteristics that indicate whether a bone is from the right or left side of the body [only for * bones].
4. Identify bone structures that constitute visible surface anatomy characteristics.

Appendicular Skeleton Questions

[use the specimens in the lab and the textbook to answer the following questions]

1. What portion of the ulna articulates with the humerus? _____

2. What part of the humerus articulates with the scapula? _____

3. What part of the femur articulates with the innominate [pelvic] bone? _____

4. What portion of the ulna articulates with the radius to allow rotation? _____

The following question will ask about structures in relation to the side of the body the structure is from. These questions are meant to point out that you will need at least two or three structures to orient a bone to the correct side of the body.

5. The prefix of "supra" indicates a structure is _____ to another structure [Use the textbook to find the meaning of prefixes]. The supraspinous fossa is larger or smaller than the infraspinous fossa? _____

Given this relationship if you were holding the right scapula in your right hand with the spine oriented towards the back of the hand [dorsum] the smaller fossa would be _____ to the larger fossa and the acromion process would curve _____ [anterior or posterior].

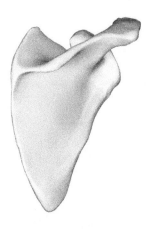

Figure 2.1 Right scapula with orientation for hand hold.

Compare how you pick up the right scapula in the right hand and how you would pick it up in the left hand. The supraspinatus fossa cannot be oriented upwards in the left hand if you are holding a right scapula. This is an example of how you should learn to hold the bones in the hand that indicates whether the bone is from the right or left side of the body.

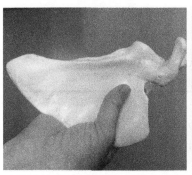 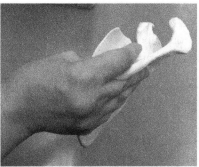 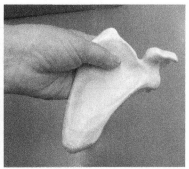

A.   B.   C.

Figure 2.2 Scapula orientations in hand to tell orientation articulated. Notice the location of the spine and which surfaces are anterior and posterior articulated and in anatomical position.

6. Which lettered diagram(s) are correctly held in the hand that is the same as their articulation on an articulated skeleton?

18 | DISSECTION SIMPLIFIED: A LAB MANUAL FOR INDEPENDENT WORK IN HUMAN ANATOMY

7. Which lettered diagram (s) are incorrectly held in their articulation hand?

8. Is the spine located on the anterior or posterior surface of the scapula (remember all questions are in reference to anatomical position).

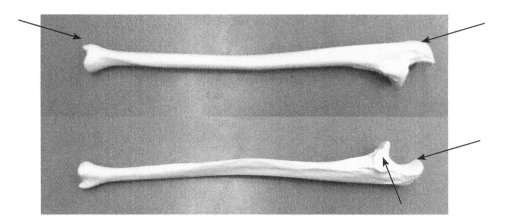

Figure 2.3 Posterior and anterior views of the ulna. Look up anatomical position to answer if the olecranon process is on the posterior or anterior surface. The ulna is easily identified by the "U" shape of the trochlear notch.

9. The bony portion of the back of the elbow is the _____.
    A. olecranon of the ulna   D. styloid process
    B. radial tuberosity   E. head of the humerus
    C. capitulum of the humerus

10. The _____ of the humerus are located at the distal end [elbow] and the _____
_____ of the humerus are located at the proximal end [shoulder].

Term bank for question 10:

capitulum trochlea	olecranon head of the humerus
greater and lesser trochanters	greater and lesser tubercles
medial and lateral epicondyles	

---

11. The bony structures the distal end of the forearm on the <u>posterior side</u>, where the forearm articulates with the wrist, are the _____ [located on the lateral and medial sides].
    A. styloid processes of the radius and ulna   C. radial tuberosity and the olecranon
    B. head of the radius and styloid process of the ulna   D. hamate and trapezium bones

12. What two bones does the clavicle articulate with?
    A. first rib and sternum
    B. sternum and humerus
    C. scapula and first rib
    D. sternum and scapula
    E. sternum and humerus

13. Which of the following structures (landmark) faces laterally?
    A. obturator foramen
    B. acetabulum
    C. greater sciatic notch
    D. ischial tuberosity
    E. iliac tuberosity

14. Hold a right pelvic bone in your right hand. If the pelvic bone is oriented as articulated on the body, the ischium will be oriented posteriorly and the pubis will be oriented anteriorly. To tell if you truly have a right pelvic bone, the acetabulum should be oriented _____.

15. If you are holding a hip bone with the acetabulum facing you, and the greater sciatic notch is to your left, you would be holding a _____ [right/left] hip bone?

16. The _____ [of the femur] are located at the _____ end of the femur [at the knee joint] and the _____ [of the tibia] are located at the _____ end of the tibia.

17. Look up the meaning of the prefix "epi" [in the textbook]; given the meaning, where are the epicondyles of the femur located in relation to the condyles? _____

18. The _____ [of the femur] are located at the _____ end of the femur, near the hip joint.

19. The bony projection at the lateral side of the ankle is the _____.
    A. lateral malleolus of the tibia
    B. medial malleolus of the tibia
    C. greater trochanter
    D. calcaneus
    E. lateral malleolus of the fibula

20. The cora_____ process of the scapula is not to be confused with the coro_____ process of the mandible.

21. The patellar surface is located on the _____ surface of the femur.

22. The meta_____ bones are found in the hand, while the meta_____ bones are found in the foot.

WORKSHEET III

Tissues and Bone Formation

Before we can learn the different tissue types in the body, we must cover the basic use of a microscope. If pictures of tissues are used for this section, students will resist learning the characteristics of the tissues (the ultimate goal of this section). Therefore, to encourage learning of tissue characteristics we will use microscope slides for tissue type identification. Following is a basic diagram of a compound microscope. Listen to the introductory lecture about the microscope and identify the arrowed parts.

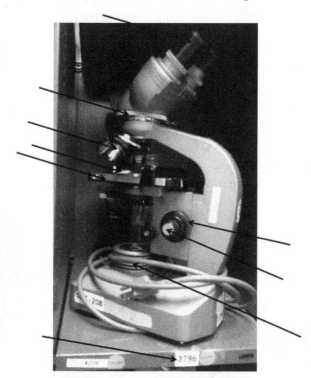

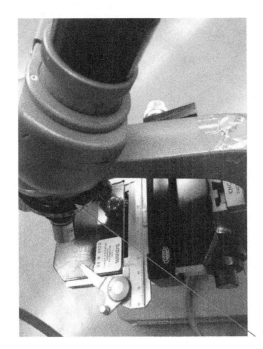

Procedure to place slide on the microscope to view tissues.

1. Place the slide on the stage and use the stage clip to secure the slide. You can then use the stage adjust knob to center the slide.

2. Turn on the light source. You can adjust the amount of light using the iris diaphragm lever; adjust the light to lower the amount if the light through the ocular is too bright.

3. Once there is a light circle, adjust the slide using the stage adjust knob so that the pink/purple tissue on the slide is in the circle of light.

4. The easiest way to focus a microscope is to raise the stage to its highest point and focus down by turning the adjustment knobs. Before raising the stage to its highest point **You Must Make Sure That the Shortest Objective is rotated into place over the slide**. Do not put the stage into the closest position unless the shortest objective is in place, it is possible to scratch the slide and lens if the objective is too long during focusing.

5. Once you have the stage adjusted to the closed distance and the sample on the slide in view; use the coarse adjust knob to focus down (or up depending on the type of microscope) while looking through the oculars.

6. You should begin to view the cells and see the tissue in focus. Now you can use the fine focus knob (the small inner knob) to refine the image resolution.

There will be several tissues in the field of view. Use your text book to help find the tissue of interest. For this example we are looking for stratified squamous epithelium. Find the tissue that looks like the

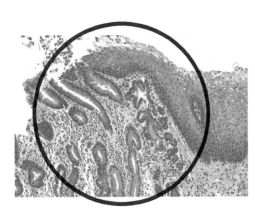
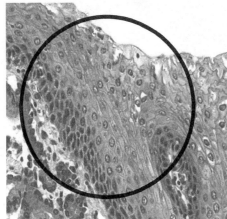
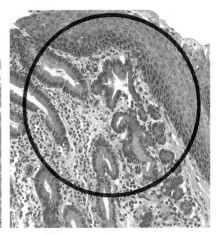

one in your text book. If it does not look like the example in your textbook it may not be the tissue you are required to find. If you have difficulties finding the tissue in the slide ask your instructor to verify you are viewing the correct example tissue (you do not want to confuse tissues while studying).

7. At 4× or 10× magnification you will not be able to view the tissue in enough detail to identify the tissue. (Tissue stations on the exam will be set-up and focused for the specific tissue to be identified. The magnification is chosen to show the necessary tissue characteristics. You are not allowed to move the slides or change the magnification yourself during and exam. If a slide is at a magnification you do not recognize the tissue you may ask the lab instructor to change the magnification. Once you have viewed the slide the lab instructor will return the slide to the original view).

8. To change magnification, please remove your eyes from the ocular area and watch as you rotate the objective to the next highest power. The microscopes have the low power next to the highest power 100× oil immersion lens. If you are not watching and rotate the 100× in place, the lens will not have enough clearance and could hit the slide on the stage. This will break the slide and scratch the lens. While looking at the objectives, rotate the next power into place (you will feel a click when the objective is in full position. The tissue of interest is in the middle of the field of view before changing magnification. The higher magnification will reduce the field of view, and you may have to move the stage (with the stage adjust knob) to view the tissue of interest.

9. Once you have found the tissue, now you need to identify characteristics that will allow recognize the tissue from the other specific tissues on the list.

The terms in bold print are the types of tissues required for the laboratory practical. The exam covering this material will require the identification of each tissue type using a microscope [there will be both slides and pictures used on the exam for this section]. Use the slides available in the lab to practice identifying the tissue types. Below the list of tissues is a flow chart you may use to organize key identification characteristics.

Tissues—

Epithelial Tissue
    Nonciliated simple columnar
    Simple ciliated columnar
    Stratified squamous

Connective Tissue
    Adipose
    Dense regular (white fibrous)
    Hyaline cartilage
    Elastic cartilage
    Fibrocartilage
    Bone—compact and spongy
    Blood
    Areolar

Muscle Tissue
    Cardiac muscle
    Skeletal muscle
    Smooth muscle

Use the Models to identify the following structures

Bone	**compact bone**	**Haversian canals**
	spongy bone	**blood vessels**
	osteocytes	**osteon**
	periosteum	**concentric Lamellae**

Activities:

1. Identify key characteristics of each tissue type
2. Identify the parts of a long bone [lecture material]
3. Identify structures on osteon model
4. Identify the steps of Intramembranous and Endochondral ossification [lecture material].

TISSUE HISTOLOGY SECTION I

Identify the characteristics that separate and classify each tissue type. For example, all epithelial tissues have an apical and basal surface.

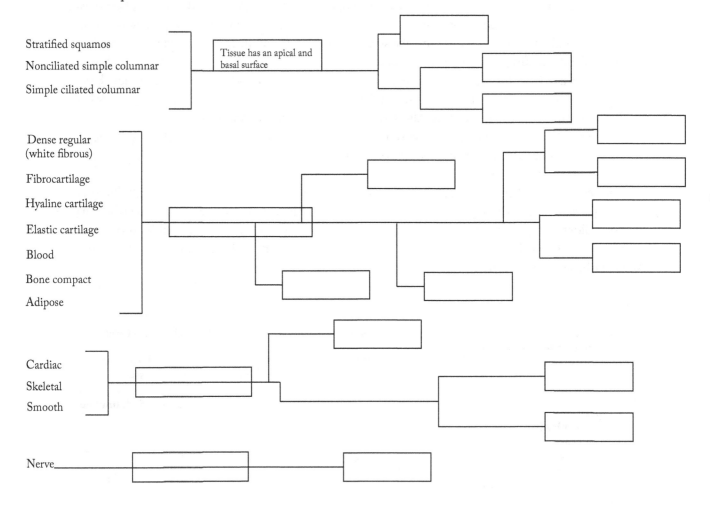

If the tissue has an apical surface

[lumen or space above—the top cells are not sandwiched between cells]

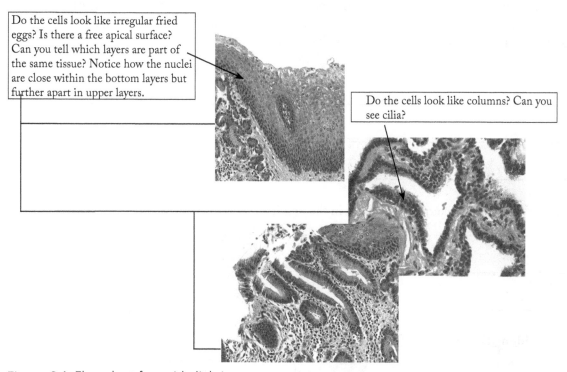

Do the cells look like irregular fried eggs? Is there a free apical surface? Can you tell which layers are part of the same tissue? Notice how the nuclei are close within the bottom layers but further apart in upper layers.

Do the cells look like columns? Can you see cilia?

Figure 3.1 Flow chart for epithelial tissue.

3.1a: Copyright © Nephron (CC BY-SA 3.0) at https://commons.wikimedia.org/wiki/File:Pancreatic_acinar_metaplasia_-_high_mag.jpg.
3.1b: Copyright © Patho (CC BY-SA 2.0) at http://commons.wikimedia.org/wiki/File:Bronchiolar_metaplasia.jpg.
3.1c: Copyright © Nephron (CC BY-SA 3.0) at https://commons.wikimedia.org/wiki/File:Pancreatic_acinar_metaplasia_-_high_mag.jpg.

Epithelial Tissue:

List five characteristics of epithelial tissue: These should be characteristics from lecture. (A good study technique is to try and list the characteristics without looking at your notes or the textbook. Write your initial characteristics in blue or black pen. Then try to look up five characteristics in your lecture notes and textbook. Write these second "looked up" characters in red, orange, or green pen. The characters you did not initially write down are the ones you do not know and need to study. I suggest this method of testing recall because I hear from students all the time "but I studied for hours". Unfortunately most students confirm information they already know when studying. Difficult or new material is ignored because it is not easily recalled. You will not notice this lack of recall until someone else asks a question about the material that does not include your known information. If you often take exams and feel too much of the material on the exam was not covered in lecture, you may have difficulty with assessing how much of the required material you know for efficient recall. Unless you have a regular study session group, this lack of recall will happen on an exam, not during study time. We often keep studying material that is comfortable because "our brains" tend to filter out unfamiliar terms. You will need study methods that test completely the information you are able to recall and know the information you cannot recall. Remember that you need to be able to recall at least 70% of the material. In a course that introduces only 30–40% new material this will seem like a reasonable amount of recalled material.

But realize if you are use to learning 70-80% of the new material, in most courses that would be only about 28% new material. The rest of the test material would be familiar 60–70%. In anatomy, unless you have taken the course before, 80–90% of the material is new. So you need to adjust to learning 56% of the new material. The large amount of new material is one of the reasons anatomy can be a challenging course. Now back to the worksheet exercise.

Epithelial Tissue characteristics

1._____   2._____

3._____   4._____

5._____

Epithelial Tissue characteristics forgotten in the first list:

1._____   2._____

3._____   4._____

5._____

How to identify Epithelial Tissues:

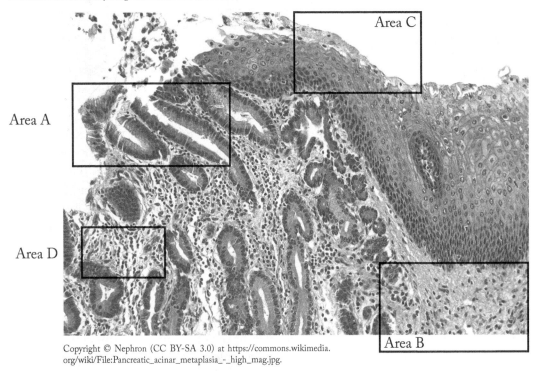

Area C

Area A

Area D

Area B

Distinguish the portions of the slide that are epithelial.

Which areas are epithelial tissues_____?

Why? What is different between the areas?_____

What is the other basic tissue type in this slide? (There are 4 basic tissue types: epithelial, connective, muscle, and nervous)_____. How is that tissue different from epithelial tissue? _____

Connective Tissue:

How to identify a connective tissue?

Connective Tissue characteristics
1._____   2._____

3._____   4._____

5._____

Connective Tissue characteristics forgotten in the first list:
1._____   2._____

3._____   4._____

5._____

Look for stained nuclei to identify individual cells. The material between is the matrix (ground substance and fibers)

Copyright © Emmanuelm (CC by 3.0) at https://commons.wikimedia.org/wiki/File:Cartilage_polarised.jpg.

Robert M. Hunt, "Hypertrophic Zone of Epiphyseal Plate," https://commons.wikimedia.org/wiki/File:Hypertrophic_Zone_of_Epiphyseal_Plate.jpg. Copyright in the Public Domain.

There are two types of connective tissues that have their cells inside a lacunae, cartilage and bone. For cartilage this gives the cells a bubble appearance. The cells in bone have the bubbled look but much smaller and harder to see.

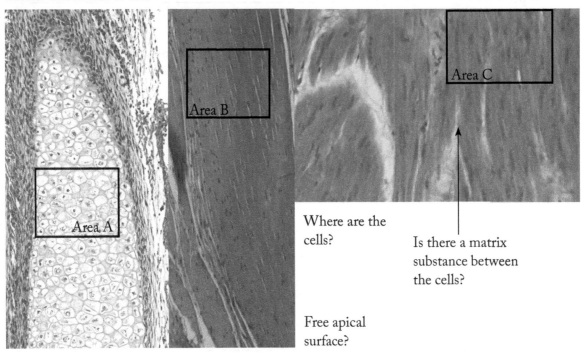

Area A

Area B

Area C

Where are the cells?

Is there a matrix substance between the cells?

Free apical surface?

Of the listed connective tissues some are easier to identify than others; What is the difference between area B and Area C? _____ Are they connective tissue or muscle tissue?

The outlines show the individual cells. Area C has individual cells that are spindle shaped. The pink stained material is contained within the cells. Area B has cells with areas of pink between the cells. So which is a muscle tissue and which is a connective tissue? _____. Notice the nuclei in Area B are more lined up together and the nuclei in area C are scatter like polka dots.

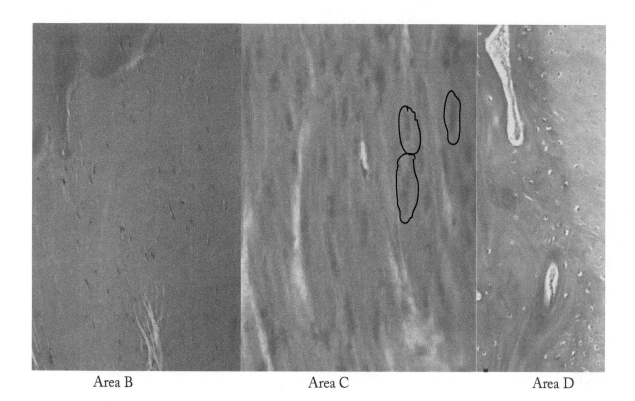

| Area B | Area C | Area D |

Area B: Copyright © Patho (CC BY-SA 3.0) at https://commons.wikimedia.org/wiki/File:Morbus_Dupuytren,_HE_7.JPG.
Area C: Copyright © Polarlys (CC BY-SA 3.0) at https://en.wikipedia.org/wiki/File:Glatte_Muskelzellen.jpg.
Area D: Copyright © Rikke K Kirk, Bente Jorgensen and Henrik E. Jensenl (CC by 2.0) at https://commons.wikimedia.org/wiki/File:Joint-lesions-sow-histo-template.jpg.

Do the cells in the background picture look different from the cells in Area B and C? Can you see a space (lacunae)? Do you see the cells as little bubbles? Also notice there are fewer nuclei per area in the background photo than in Area C.

Do the nuclei look different in area E?_____
Is area E muscle or connective tissue?_____

There are some tissues that are more difficult to identify and some that are easily confused. The previous page illustrated several tissues that are easily confused. The most commonly confused tissues are dense regular connective and smooth muscle.

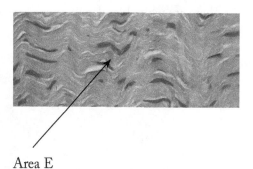

Area E

Copyright © Patho (CC BY-SA 3.0) at https://commons.wikimedia.org/wiki/File:Morbus_Dupuytren,_HE_6.JPG.

If you do not notice the lacunae in fibrocartilage, it can look like dense regular. The nuclei in dense regular connective tissue tend to be flattened by the collagen fibers and are not as round as those in smooth muscle or fibrocartilage.

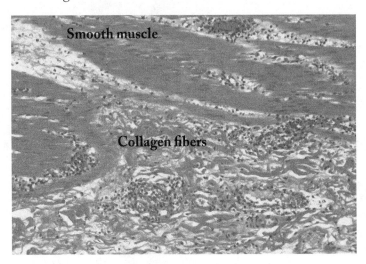

Figure A

In many cases you will have to view the two tissues side by side to find a difference that could be used to identify each tissue correctly. Using color is not a good character since the color is due to the staining procedure (notice that hyaline cartilage slides can appear pink or purple. However, looking for common pattern differences can help. The nuclei of smooth muscle are staggered because the cells are spindled shape. When looking under a low power, the nuclei appear as polka dots scattered in a sea of collagen. The nuclei of the dense regular tissue are smashed between extra cellular matrix and are not scattered. The nuclei tend to line up and are less dense in comparison to the collagen. This difference in pattern can give you a starting point and viewing several examples allows you to test your identification skills.

There are several connective tissues on the list that do not require much explanation. Blood is easy to identify, since the erythrocytes are anucleated.

There are white blood cells in the specimen but you will not have to identify specific cells until we study the circulatory system.

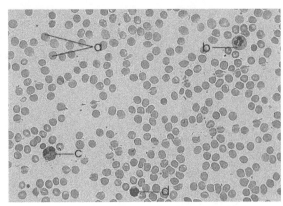

Another tissue that is distinctive is adipose. This connective tissue has little extra cellular matrix (which is different from the other connective tissues. Adipose can look similar to bone marrow, but bone marrow is not on the class list. However, spongy bone is on the class list so bone marrow will be in the field of view for a spongy bone specimen. Compare the two and notice the number of blood cells in the marrow.

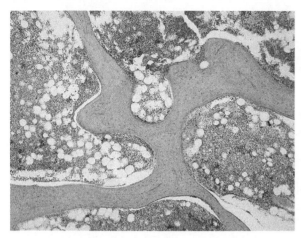 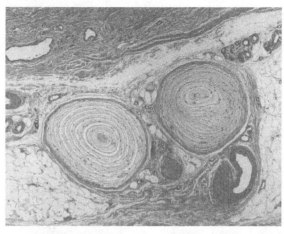

Figure B. Spongy bone with trabeculae and bone marrow. Dermis with adipose tissue. There is also smooth muscle in the slide; can you find it.?

B1: Copyright © Reytan (CC BY-SA 3.0) at https://commons.wikimedia.org/wiki/File:Spongy_bone_-_trabecules.jpg.
B2: Copyright © Patho (CC BY-SA 3.0) at https://commons.wikimedia.org/wiki/File:Lamellar_corpuscle,_HE.JPG.

Muscle tissue does not have an extra cellular matrix but it can be difficult to see if the material is inside the cell or outside the cell. Striations (alternating lighter and darker bands) is characteristic of two of the tissues, but smooth muscle does not have any striations so this characteristic cannot be used as a unique character for identifying muscle tissue. If there are striations in the tissue there are two possibilities, skeletal or cardiac muscle. Skeletal muscle is multinucleated and does not branch. Cardiac muscle is uninucleated, branched, and has intercalated discs. The intercalated discs indicate the ends of the cell, which allows many of the cells to be distinguished as uninucleated. Review the slides in Figure C and identify the skeletal muscle from the cardiac muscle tissue.

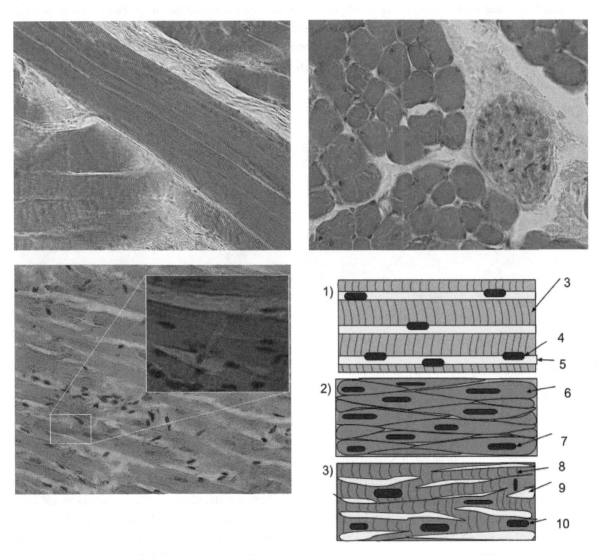

Figure C. Longitudinal sections of cardiac and skeletal muscle tissue. There is a cross section of skeletal muscle. This slide is included as an example of what portions of the specimen not to use for identification. The characteristics for identification are different for a cross section. Therefore the specimens used for the lab exams will be longitudinal sections, not cross sections of the organs. The illustration shows diagrams of the three muscle types, match the number to the type of muscle tissue.

<u>If the tissue has a visible matrix</u>

[extracellular non-living substances]

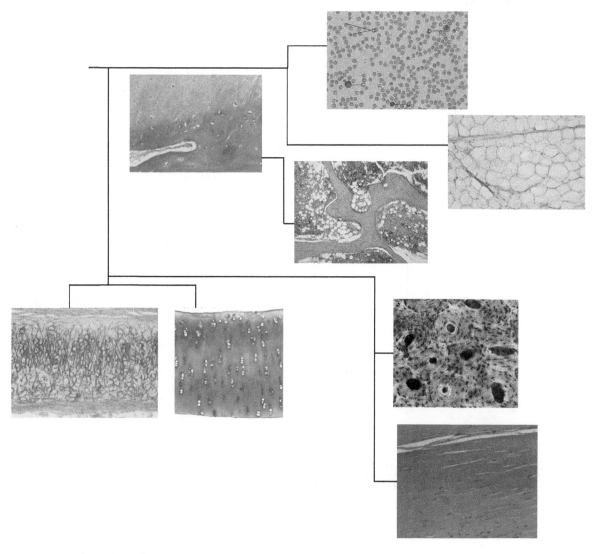

Figure 3.2 Flow chart for connective tissues.

3.2a: Adapted from: Copyright © Rikke K Kirk, Bente Jorgensen and Henrik E. Jensenl (CC by 2.0) at: https://commons.wikimedia.org/wiki/File:Joint-lesions-sow-histo-template.jpg.

3.2b: Copyright © Andrea Mazza (CC BY-SA 3.0) at https://commons.wikimedia.org/wiki/File:Cartilago_elastico_100X.JPG.

3.2c: Copyright © Takeshi Teramura, Kanji Fukuda, Shinji Kurashimo, Yoshihiko Hosoi, Yoshihisa Miki, Shigeki Asada and Chiaki Hamanishi (CC by 2.0) at https://commons.wikimedia.org/wiki/File:Bovine_cartilage_tol_blue_and_alcian_blue.jpeg.

3.2d: Copyright © Reytan (CC by 3.0) at http://commons.wikimedia.org/wiki/File:Blood_smear.jpg.

3.2e: Copyright © Reytan (CC BY-SA 3.0) at http://commons.wikimedia.org/wiki/File:Yellow_adipose_tissue_in_paraffin_section_-_lipids_washed_out.jpg.

3.2f: Copyright © Reytan (CC BY-SA 3.0) at http://commons.wikimedia.org/wiki/File:Spongy_bone_-_trabecules.jpg.

3.2g: Copyright © Reytan (CC BY-SA 3.0) at http://commons.wikimedia.org/wiki/File:Compact_bone_-_ground_cross_section.jpg.

<u>If the tissues does not have a visible matrix</u>

[cells next to cells ... look for the membrane lines] and no apical surface-layers of cells

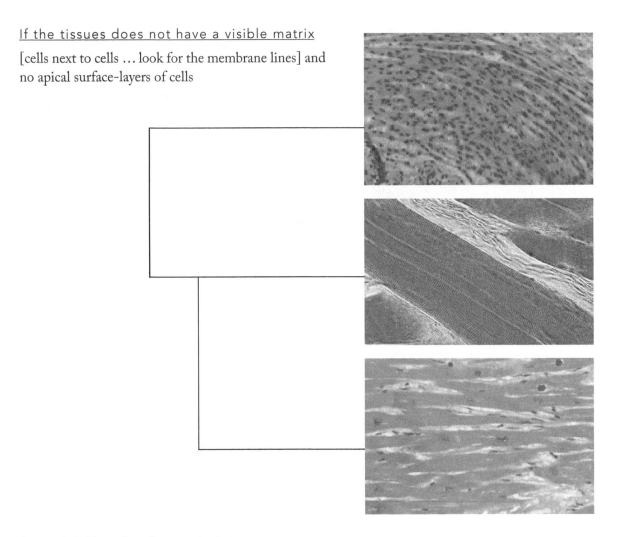

Figure 3.3 Flow chart for muscle tissues.

TISSUE IDENTIFICATION

<u>Tissue Identification A</u>

Identify the two tissues and explain differences between them.

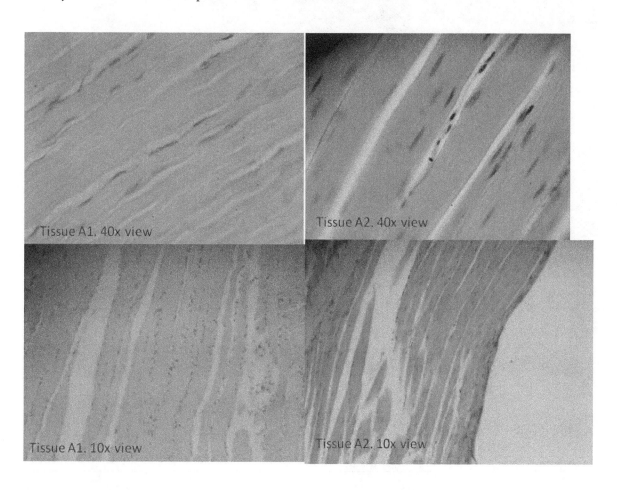

1. Identify Tissue A1 _____

2. Identify Tissue A2 _____

How are Tissue A1 and A2 different? _____

Similarities? _____

Tissue Identification B.

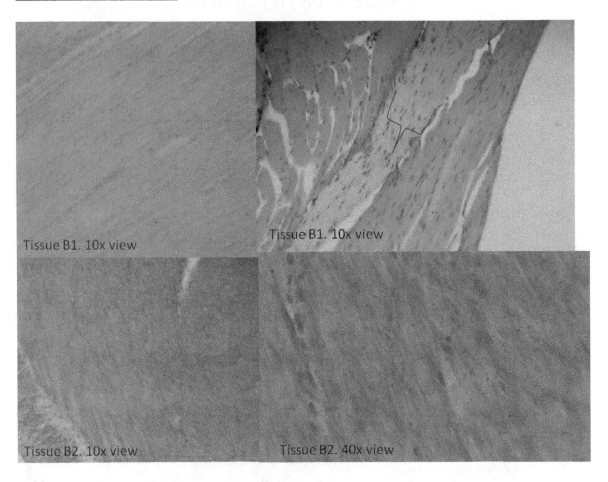

Tissue B1. 10x view

Tissue B1. 10x view

Tissue B2. 10x view

Tissue B2. 40x view

3. Identify Tissue B1 _____

4. Identify Tissue B2 _____

How are Tissue B1 and B2 different? _____

Similarities? _____

Tissue Identification C.

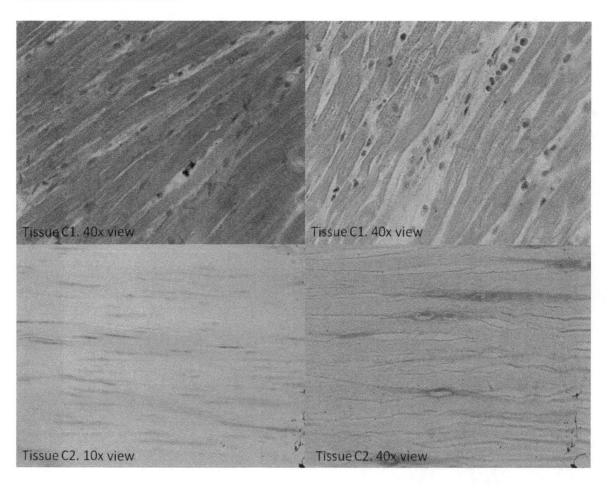

5. Identify Tissue C1 _____

6. Identify Tissue C2 _____

How are Tissue C1 and C2 different?_____

Similarities?_____

Tissue Identification D.

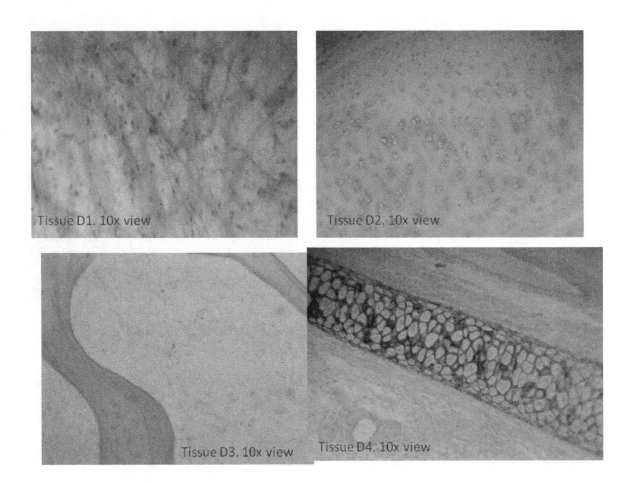

Tissue D1. 10x view

Tissue D2. 10x view

Tissue D3. 10x view

Tissue D4. 10x view

7. Identify Tissue D1 _____

8. Identify Tissue D2 _____

9. Identify Tissue D3 _____

10. Identify Tissue D4 _____

BONE STRUCTURE AND GROWTH SECTION II

<u>Anatomy of a long bone</u>

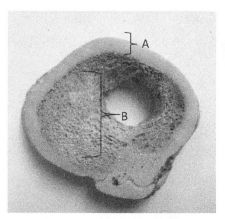

Figure A

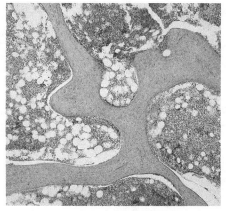

This tissue slide illustrates bone that would be found in which section of the long bone 7, 8, or 9 ?
_____

Copyright © Reytan (CC BY-SA 3.0) at https://commons.wikimedia.org/wiki/File:Spongy_bone_-_trabecules.jpg.

<u>Questions for Anatomy of a Long Bone:</u>

Identify structures: 1–10:

1. _____ [type of tissue]
2. _____ [what is found within the bone type in question 1 in area 2?].
3. _____ [type of bone tissue].
4. _____
5. _____
6. _____
7. _____
8. _____
9. _____
10. _____

Region A [refer to Figure A] is composed of _____bone and region B is composed of _____ bone.

*Note the identified structure #3 is the same answer as region B

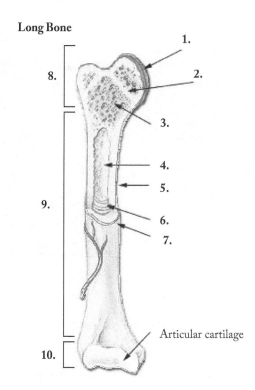

Long Bone

Articular cartilage

Figure 3.4 Anatomy of a long bone.

National Cancer Institute, "Illu long bone," https://commons.wikimedia.org/wiki/File:Illu_long_bone.jpg. Copyright in the Public Domain.

Histology of Bone Tissue

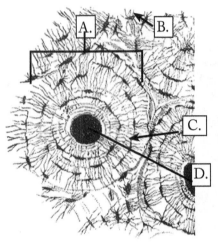

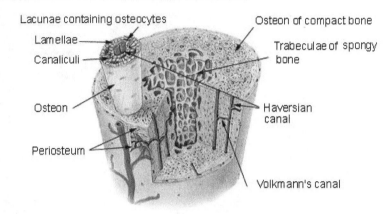

Figure 3.5 Bone histology.

Henry Vandyke Carter, "Transverse section of compact tissue bone," http://commons.wikimedia. org/wiki/File:Gray73.png. Copyright in the Public Domain.

Figure 3.6 Illustrated and labeled bone histology.

National Cancer Institute, "Compacy spongy bone," https://commons.wikimedia.org/wiki/ File:Illu_compact_spongy_bone.jpg. Copyright in the Public Domain.

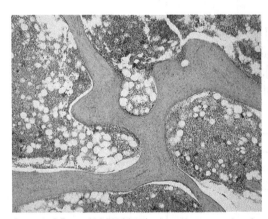

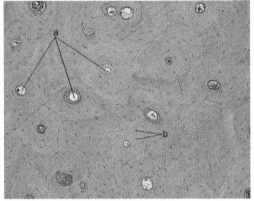

Copyright © Reytan (CC BY-SA 3.0) at https://commons.wikimedia. org/wiki/File:Spongy_bone_-_trabecules.jpg.

Copyright © Reytan (CC BY-SA 3.0) at http://commons.wikime-dia.org/wiki/File:Compact_bone_-_decalcified_cross_section.jpg.

Questions for Histology of Bone:

11. Identify A–D
12. View the slides of spongy and compact bone: What histological structure does compact bone have that is not found in spongy bone? _____
13. What structure is visible within spongy bone that is not present in compact bone?

_____

Bone Formation

Intramembranous Ossification (refer to the lecture text on the section of bone formation)

Step 1: Development of the center of ossification
What types of cells cluster together at the center of ossification in Intramembranous ossification?

_____

If a bone matrix is formed, what type of cell must these cells differentiate into? _____

Step 2: Calcification
The _____ cells secrete the fibrous portion of the bone matrix. What type of fiber composes most of the extracellular matrix before calcium and other mineral salts are deposited? _____ When the extracellular matrix formation stops, the cells inside the lacunae are _____.

Step 3: Formation of trabeculae
When the blood vessels grow into spaces formed by the extracellular matrix composing the trabeculae, the type of bone tissue formed is _____ [compact or spongy]. As the bone tissue develops, the _____ cells condense on the periphery to form the _____ _____, [which is step 4 of the bone formation process].

Endochondral Ossification

Step 1: _____

What type of cells do mesenchyme cells differentiate into in the first step of endochondral ossification?

_____

Step 2: What type of growth results from cell division of the chondrocyte and secretion of the extracellular matrix? _____

Step 2 cont.: Appositional growth results from the addition of _____ and will increase the developing cartilage model in length or width? _____

Step 3: The primary ossification center is initiated by _____ _____ and is located at the _____ [ends, middle, top, bottom] of the cartilage model.

Step 4: The secondary ossification center develops in the _____ [epiphysis or diaphysis].

The _____ is located between the diaphysis and epiphysis, which will be responsible for lengthwise growth.

In the figure below, identify the important changes in each step of ossification:

1. _____

2. _____

3. _____

4. _____

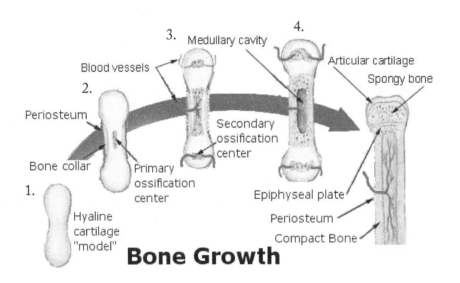

Figure 3. 7 Endochondral bone ossification.

Chaldor, "endochondral bone ossification," https://commons.wikimedia.org/wiki/File:Bone_growth.png. Copyright in the Public Domain.

WORKSHEET IV

Skeletal Joints

Use the Joint models available in the lab to identify the terms in bold print. This worksheet contains questions that will aid studying for lecture exam questions as well as lab practical questions.

Knee joint:	tibial collateral ligament	medial meniscus
	fibular collateral ligament	lateral meniscus
	patellar ligament	quadriceps tendon
	anterior and posterior cruciate ligaments	

Identify which ligaments in the knee joint are located outside the **articular capsule** and which ligaments are located inside the articular capsule [note the articular capsule of the knee is not a complete independent capsule].

Outside ligaments [Extracapsular ligaments]   1. _____

    2. _____

Inside ligaments [Intracapsular ligaments]   1. _____

    2. _____

Compare the models in the lab to the illustrations.
Is the articular capsule visible in the models available in the laboratory? _____

What is the function of the medial and lateral menisci?
_____

_____

Are these menisci located inside or outside of the Synovial cavity? _____

The Synovial cavity is colored blue: review the colored original picture

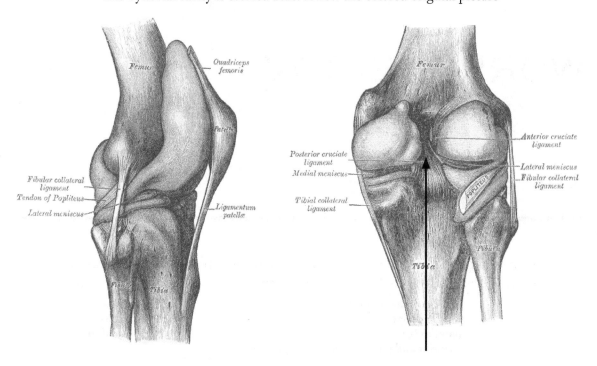

Figure 4.1 Posterior view of the knee joint; note the location of the posterior cruciate ligament. This illustration depicts the knee joint's synovial lining membrane, diagrammatically. This tissue layer lies just inside of the joint's surrounding capsular (envelope) ligament, and encloses the internal joint space. The picture depicts the synovial membrane as if it were distended with fluid, as often occurs in PVNS disease, thus accounting for its expanded appearance.

4.1a: Henry Gray, "Posterior view of the knee joint," http://commons.wikimedia.org/wiki/File:Gray351.png. Copyright in the Public Domain.
4.1b: Henry Gray, "Posterior view of the knee joint," http://commons.wikimedia.org/wiki/File:Gray352.png. Copyright in the Public Domain.

To aid in remembering which ligaments are lateral and which are medial, answer the following question: Which long bone on an articulated skeleton is located most laterally to the knee joint?

_____

TYPES OF SYNOVIAL JOINTS

Planar Joints

1. Type of movement? _____

2. Identify two examples: _____

_____

Use the model of the right elbow to answer the following questions.

3. The articulation between the ulna and humerus forms which type of synovial joint?

    _____

4. Which type of synovial joint allows the digits to move biaxially? [the digits and move anterior-posterior and medial-lateral] _____?

5. The head of the radius articulates with the _____ of the ulna. The articulation is an example of a _____ joint.

Review the shoulder model and be able to identify the skeletal components.

6. The entire upper limb has its bony attachment to the axial skeleton at what joint? _____ [hint refer to the text book for a definition of the structures that compose the upper limb]
    A. glenohumeral joint
    B. acromioclavicular joint
    C. sternoclavicular joint
    D. scapulovertebral joint

    The answer in question 6 is which type of Synovial joint? _____

Review the hip muscle model to answer the following questions. Also as a review of the skeletal system, you should be able to identify all skeletal components on the model.

7. The articular capsule encompasses which of the following femoral structures [circle all that are correct]

    greater trochanter   lesser trochanter   lateral condyle

    head of femur   linea aspera   medial condyle

8. The hip joint is an example of which type of synovial joint?

    _____

9. The ligaments that compose the articular capsule are composed of mostly which type of connective tissue? _____ _____ [hint: this is a review of the tissue lab; refer to the tissue textbook chapter]

10. Which type of connection joins the gluteus medius muscle to the greater trochanter of the femur? _____ [hint: this is the general term for a structure that connects muscle to bone]

11. All synovial joints are classified <u>functionally</u> as? _____ _____
    A. Synarthorsis
    B. Amphiarthrosis
    C. Diarthrosis
    D. Fibrous

Psoas major
Iliacus
Tensor fasciae latae
Piriformis
Adductor brevis
Adductor longus
Pectineus
Iliotibial tract
Gracilis
Adductor magnus

The functional classifications of joints—this is how much movement the joint allows. Remember that all synovial joints are a type of diarthrosis, but that cartilaginous and fibrous can be either amphiarthrosis or synarthrosis.

To remember the three functional classifications, use the meaning of the prefix and suffix [use the glossary in the back of the textbook for prefix and suffix meanings]

the suffix "arthro" means = _____

To remember the suffix in the terms for functional classification of joints, associate the term with a common disease that is painful and limits the movement of joints _____.

Syn = _____: develop an association with the meaning to remember which types of joints are classified as synarthorsis.

Amphi = both [this one is not in the glossary]

Structural classification of joints:

12. The sutures between the cranial bones are examples of what kind of structural joint?

_____

13. The suture joints are also **functionally** classified as _____ since there is no movement of the bones around the joint.

14. The joint between the tibia and fibula is only slightly movable without a synovial membrane and is an example of which kind of **structural** joint? _____

15. The intervertebral discs are composed of [type of connective tissue] _____ and form a _____ joint [**structural classification**] between the vertebral bodies of the vertebrae.

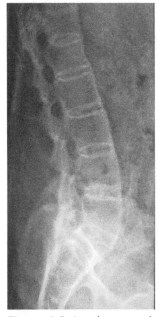

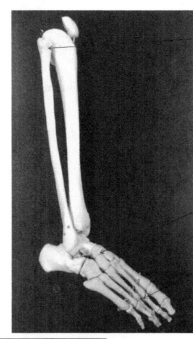

16. Structural joint between tibia and fibula?_____

17. functional joint between tibia and fibula?_____

18. Type of synovial joint?_____

19. Type of synovial joint?_____

Figure 4.2. Lumbar vertebrae.

Figure 4.3. Lower limb joints

Figure 4.2: Overkilled, "Lumbar vertebrae," https://commons.wikimedia.org/wiki/File:Ankylosing_spondylitis_lumbar_spine.jpg. Copyright in the Public Domain.

WORKSHEET V

Muscles of the Thigh, Leg, and Hip

Dissect and identify the following muscles in bold print. Know the origin, insertion, and action of muscles with an *. Be able to identify the muscles on both cat specimens and human muscle models.

Identify the muscles based upon their locations on the cat. [Note that since humans are bipeds, some of the leg muscles are oriented differently from the same muscle on a cat.]

List of Muscles

Medial Thigh
  sartorius
  gracilis
  ***rectus femoris**
  ***vastus medialis**
  ***adductor longus**
  adductor femoris (cat only)
  ***adductor magnus and brevis in the human models only**
  ***semimembranosus**
  ***semitendinosus**

Lateral Thigh

  ***biceps femoris**
  ***vastus lateralis**
  ***vastus intermedius**

Anterior Leg
  ***tibialis anterior [cranialis]**

Posterior Leg
  ***gastrocnemius**
  ***soleus**

Hip [lateral]

  ***tensor fasciae latae**
  ***gluteus maximus**
  ***gluteus medius**
  iliopsoas [on the cat]

Identify on the human model of the hip:
  ***iliacus**
  ***psoas major**

Medial Leg
  Flexor digitorium longus

Lateral Leg
  ***extensor digitorum longus**
  peroneus group (fibularis group)

47

Activities

1. Begin the dissection of the cat, skin the hind limb, hip, and back to the center of the spine.
2. Identify and dissect [clear fat and connective tissue from muscle surfaces] the listed muscles on the cat.

CAT DISSECTION

Step 1. To begin the dissection use the blunt-tipped scissors to make an incision in the lower abdominal.

[Notice that the skin is loose in the lower region, so you will be able to cut the skin without cutting through the muscle tissue, which would open the peritoneal cavity.]

Step 2. The skin has to be separated from the underlying muscle tissue. The most effective method of separating the skin is by clearing away the connective tissue with your finger or using a blunt probe [do not use a sharp probe or blindly cut the skin without separating the skin from the underlying tissue]. The skin has to be separated from the hind limb muscles but you should follow the illustrated pattern so that the skin remains attached at a few points and can be folded back down on the cat. Leaving the skin attached keeps the muscles from drying out.

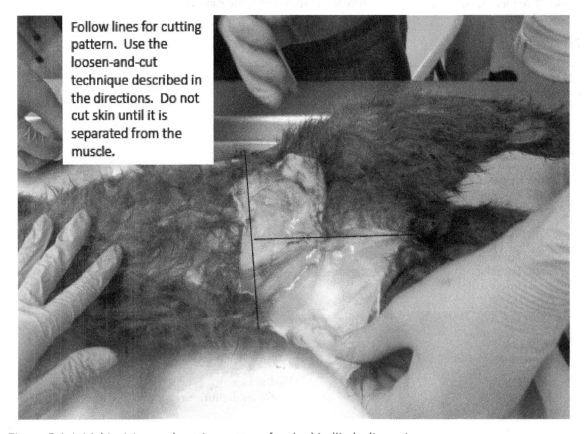

Follow lines for cutting pattern. Use the loosen-and-cut technique described in the directions. Do not cut skin until it is separated from the muscle.

Figure 5.1 Initial incision and cutting pattern for the hindlimb dissection.

Separating the skin from the muscle tissue can be confusing if you are unfamiliar with the appearance of muscle, fat, and connective tissue. The muscles are bundled together and tend to remain on the body when the skin is separated slowly. Most of the loose, easily separated tissue is fat [yellowish to white in color] and connective tissue [white fibers] attached to the skin and the surface of the muscles.

Step 3: The skin has to be separated from the underlying muscle tissue. The most effective method of separating the skin is by clearing away the connective tissue with your finger or using a blunt probe [do not use a sharp probe or blindly cut the skin without separating the skin from the underlying tissue]. The skin has to be separated from the hind limb muscles but you should follow the illustrated pattern so that the skin remains attached at a few points and can be folded back down on the cat. Leaving the skin attached keeps the muscles from drying out.

 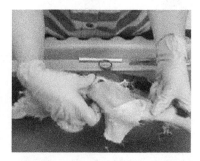

Separating the skin from the muscle tissue can be confusing if you are unfamiliar with the appearance of muscle, fat, and connective tissue. The muscles are bundled together and tend to remain on the body when the skin is separated slowly. Most of the loose, easily separated tissue is fat [yellowish to white in color] and connective tissue [white fibers] attached to the skin and the surface of the muscles. Before blindly cutting review the dissection techniques to practice during the hind limb dissection.

Techniques

There are several dissection techniques that will be useful to learn while skinning the lower leg. Following are techniques you need to practice because they will be used during later dissections of the upper body.

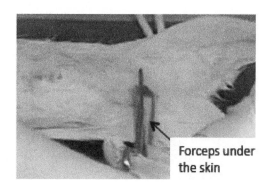

Forceps under the skin

1. Using forceps to protect underlying muscle tissue during skinning. Use the blunt forceps to loosen the skin from the underlying tissue. Muscle tissue is bundled together with connective tissue and is attached to the skin by areolar connective tissue. Areolar connective tissue is loose and spidery. The fibrous connective tissue that bundles muscles together is tougher and not easily loosened with forceps. This difference allows you to separate the skin from the muscle tissue before cutting the skin (caution if you cut the skin without loosening, you will tend to cut and damage the underlying muscle). So as you can see in the illustration, loosen the skin tissue and then cut.

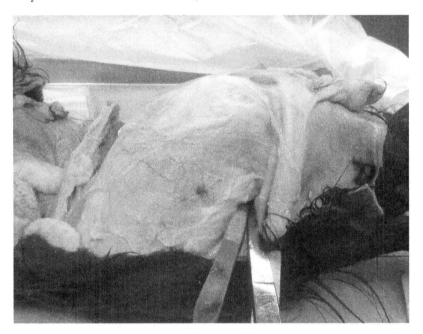

2. While skinning there will be areas around "corners" that are difficult to separate the skin (the surface will not be flat). To prevent cutting underlying muscles use the blunt probe to push through the areolar connective tissue. After pushing the probe through, guide the scissors down the probe shaft (allowing the probe to protect the underlying muscle) and cut along the probe metal. Try this technique on the posterior side of the lower thigh to prevent cutting the semitendinosus (this technique will also be needed in the upper arm to prevent cutting the pectoral and brachial muscles).

3. While skinning the cat, the skin will not easily pull away from the muscle tissue. To separate the skin from the underlying muscle tissue use the following pull and clip method.

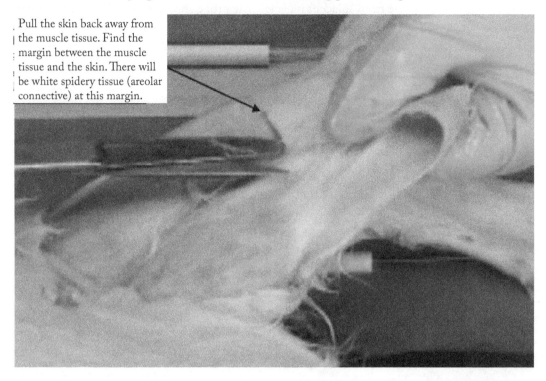

Pull the skin back away from the muscle tissue. Find the margin between the muscle tissue and the skin. There will be white spidery tissue (areolar connective) at this margin.

Pull the skin away from the underlying muscle tissue. Find the margin between the skin and muscle tissue. Use the scissors, with blunt tip nearer to the muscle tissue, to clip upwards at the margin between the two tissues. As you are clipping you can pull on the skin so that it loosens from the muscle tissue. The scissors should be angled upwards and always clip towards the skin. This will keep the scissors from damaging the underlying muscle.

4. To view deep muscles you will have to bisect superficial muscles. The superficial muscles are bisected to preserve their origin and insertion points, which allow the muscles to be correctly identified even though they are cut. To correctly bisect a muscle you will have to use the blunt probe to isolate the outer margins of the muscle.

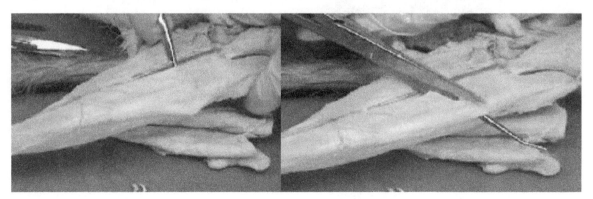

Use the images in the lab manual to identify the shape and edges of the muscle. Identifying the origin and insertion will also help with finding the margins of the target muscle. Use the blunt probe to loosen the muscle from adjacent muscles. The probe should easily clear the loose connective tissue but should not tear the bundled muscle. Slide the probe under the muscle to be bisected and push through to the opposite margin. Leave the probe under the muscle and cut along the probe (allow the probe to protect the deep muscles).

Once the muscle is bisected you will need to reflect both halves back to view the deep muscles.

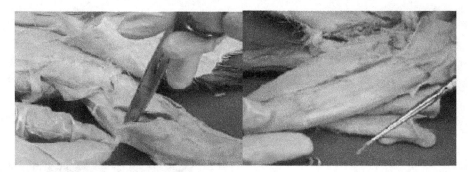

5. To view individual muscles you will have to separate the muscle bundles by clearing the connective tissue holding the bundles together.

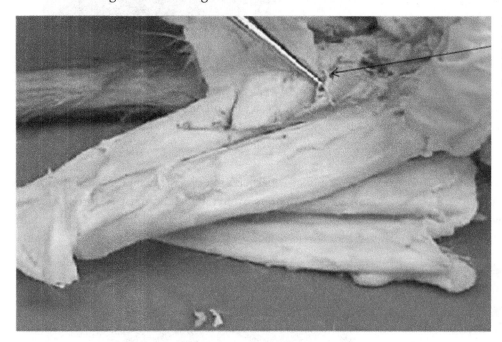

Isolate the muscles without peeling fibers away from the muscles.

The blunt probe should clear the loose connective tissue easily without tearing the muscle bundles. The lab illustrations should also help with finding the margins between muscles (looking at pictures of already separated should allow you to find the margins on a non-dissected cat).

6. Muscle tissue is bundled with fibrous connective tissue that does not separate as easily as areolar connective tissue. Cats will vary in the amount of connective tissue (male cats will often have more and stronger dense connective tissue around muscles). When the connective tissue is more dense you will have to use scissors to cut the tissue.

An example of when this will be required is while separating the tensor faciae latae from the gluteus maximus.

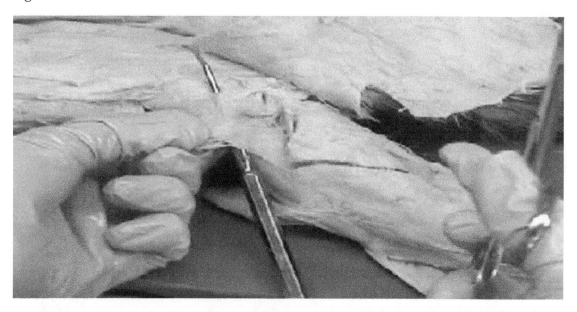

Connective tissue is lighter in color and thinner than muscle tissue. Use the blunt probe to slide under the connective tissue area and then cut where the tissue is less opaque and does not contain muscle fibers.

Tensor fasciae latae pulled back

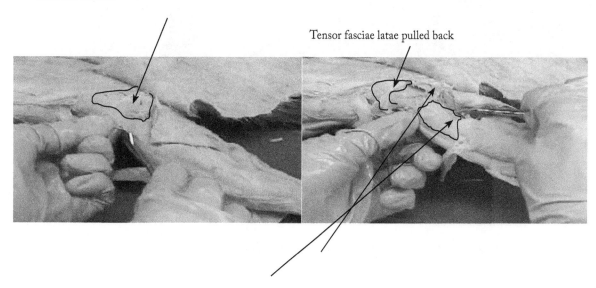

The tensor fasciae latae is connected to the gluteus maximus. When cut apart now the gluteus maximus is in view and the vastus lateralis.

Once the connective tissue is cut you can use the blunt probe to loosen the muscles.

Dissect medial superficial thigh muscles

sartorius
gracilis

Step 1. Locate the vein running down the middle of the medial thigh. At this point there is a separation between the two medial superficial thigh muscles.

Step 2. Use a blunt probe and slide the probe under the thin superior medial thigh muscle. Gently sliding the probe under the muscle, loosen the tissue underneath and work the probe to the other side of the muscle so that it emerges through the connective tissue on the lateral edge of the sartorius. Once you have both margins visible, use the scissors to bisect the sartorius muscle.

Step 3. Repeat this procedure to bisect the gracilis [inferior superficial medial thigh muscle]. Be careful when bisecting the gracilis that you do not cut the semitendinosus.

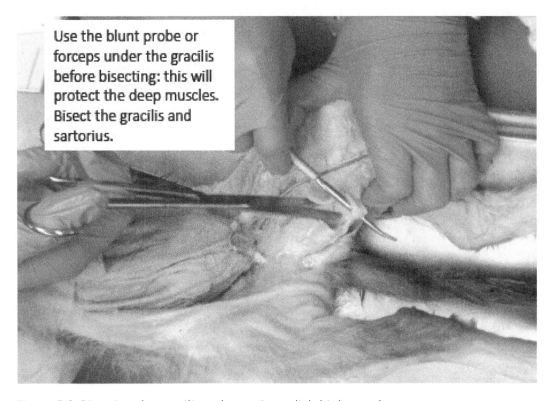

Use the blunt probe or forceps under the gracilis before bisecting: this will protect the deep muscles. Bisect the gracilis and sartorius.

Figure 5.2 Bisecting the gracilis and sartoris medial thigh muscles.

Step 4. Once the sartorius and gracilis have been bisected, the deep muscles have to be separated and defined by clearing away any fascia or fat covering the muscles. [use the blunt forceps and probe to clear fascia away.... any sharp tools will tear the muscle tissue]

Step 5. Locate the deep medial thigh muscles under the bisected gracilis.

These muscles are arranged side by side so it is helpful to find the first one in the series. The adductor longus is the most anterior of this group of muscles.	adductor longus
	adductor femoris (cat only)
	semimembranosus
	semitendinosus

This adductor longus is separated. You need to clear connective tissue to see the same on the right leg in this picture.

Use the blunt probe or forceps to separate muscles

Locate the femoral vein.

Figure 5.3 Clearing the connective tissue from the deep muscles under the gracilis.

Step 6. The next muscle in the medial row is the adductor femoris. [Be careful separating this muscle because it has two heads.... Carefully find the margin between the adductor femoris and the

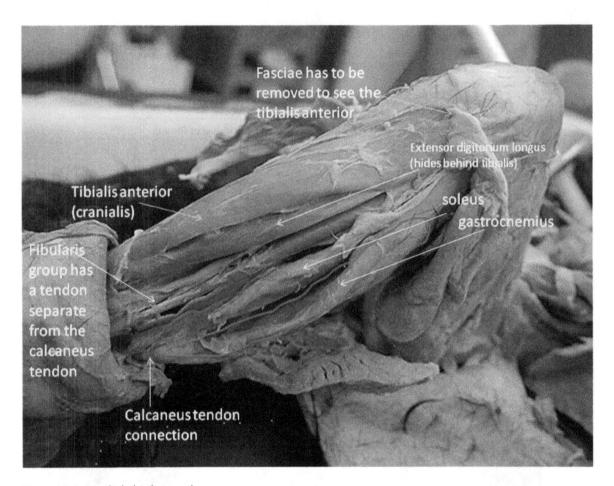

Figure 5.4 Medial thigh muscles.

semimembranosus…. If you have to use scissors to cut the tissue to separate the muscle from the other muscles—Stop—you will be separating one head of the adductor femoris from the other head.] Muscles separate from each other more easily than the fibers of individual muscles separate … therefore you should be able to separate most of the muscles using only the blunt probe or forceps … cut only the white connective tissue to expose the muscle surface.

Step 7. The next muscle is the semimembranosus, which should be separated from the final muscle in the series the semitendinosus. The semitendinosus is the posterior muscle and has the characteristic shape of the hamstring muscle group. [Cats have a pocket of fat attached to the posterior portion of the semitendinosus. This fat pocket contains the anal gland and can be removed.]

Step 8. Completely separating the semitendinosus should be left until the biceps femoris has been bisected on the lateral side of the thigh.

Step 9. Dissect muscles superior to the femoral vein and deep to the sartorius.

iliopsoas [on the cat]
rectus femoris
vastus medialis

Use the forceps to clear the connective tissue away from the muscles to expose each muscle shape. The iliopsoas is located under the femoral vein so you will have to cut the vein mid-way across the femur and reflect it back before the iliopsoas is visible. The vastus medialis is the first muscle parallel to the femur superior to the femoral vein. The rectus femoris is the just superior to the vastus medialis.

Step 10. Locate the tensor fasciae latae to begin the dissection of the lateral thigh.

Lateral thigh		Hip [lateral]	
	biceps femoris		tensor fasciae latae
	vastus lateralis		gluteus maximus
	vastus intermedius		gluteus medius

The lateral thigh will require the bisection of both the tensor fasciae latae and the biceps femoris.

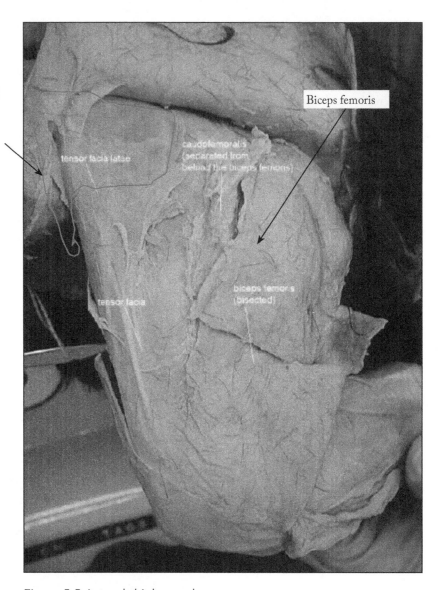

Figure 5.5 Lateral thigh muscles.

Step 11. If you are unable to locate the triangular shape of the tensor fasciae latae, then locate the fasciae connecting this muscle to the patellar region.

Use a blunt probe to loosen the connective tissue and bisect through the fasciae latae ... this bisection will allow the tensor fasciae latae to be lifted up and to expose the deep muscles.

Step 12. Use the blunt probe to locate the margins of the biceps femoris [be careful not to include the sciatic nerve located underneath the biceps femoris]. While using the probe to protect the underlying muscle, cut the biceps femoris.

- Note: at this time you should locate the semitendinosus so that it is not cut in half with the biceps femoris.... see the next page for the location of the semitendinosus.

Step 13. Separate the gluteus medius and gluteus maximus.

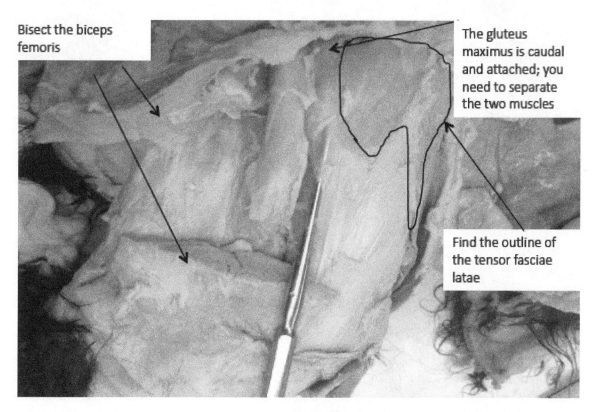

Figure 5.6 Separating the tensor fasciae latae and gluteus maximus.

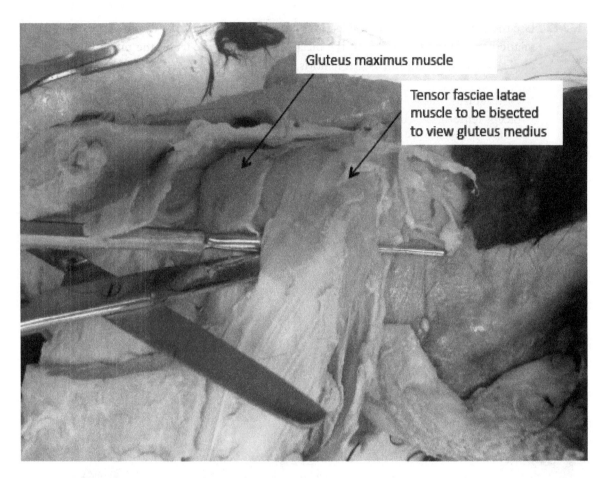

Gluteus maximus muscle

Tensor fasciae latae muscle to be bisected to view gluteus medius

Figure 5.7 Bisecting the tensor fasciae latae to view the gluteus medius muscle.

Step 14. The gluteus maximus on the cat is not as large as it is on the human. To find this muscle locate the biceps femoris anterior margin … the next muscle in line is the caudofemoralis [which is a muscle you do not have to know but has to be separated to locate the gluteus maximus]. The next muscle after the caudofemoralis is the gluteus maximus … this muscle will not always be visible until after the tensor fasciae latae has been peeled back.

Step 15. The gluteus medius is the next muscle anterior to the gluteus maximus. This muscle is larger and has a rounded margin on the lateral side.

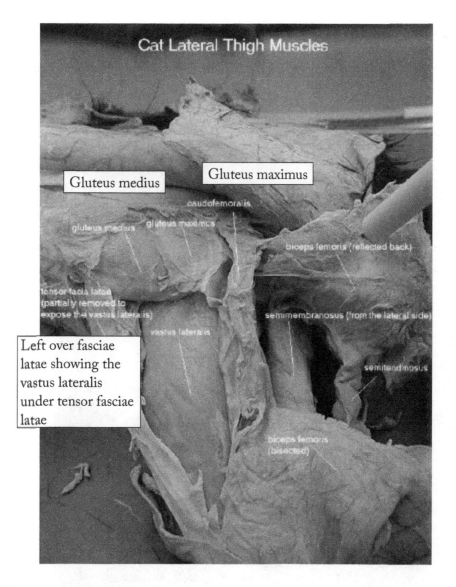

Figure 5.8 Deep lateral thigh muscles.

Step 16. The vastus lateralis is located on the lateral thigh and is visible once the fasciae latae has been peeled back.

The vastus intermedius is a deep muscle that is included because it is the fourth muscle of the quadriceps femoris. This muscle is difficult to locate ... and is found between the vastus lateralis and the rectus femoris.

Step 17. Dissect and separate lower leg muscles.

These muscles require the connective tissue to be separated to reveal the margins of each muscle. The first margin to define is the anterior surface [superficial] muscle.

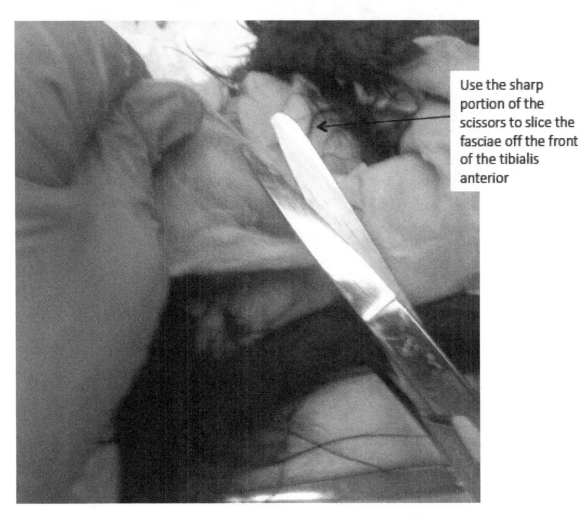

Use the sharp portion of the scissors to slice the fasciae off the front of the tibialis anterior

Figure 5.9 Expose the tibialis anterior.

Anterior Leg	
	tibialis anterior [cranialis]

Step 18. The Posterior superficial muscle is the gastrocnemius ... this muscle has two heads and can be seen from both the lateral and medial side.

Posterior Leg	Medial Leg
gastrocnemius	flexors of foot [known as a group]
soleus	
	Lateral Leg
	extensor digitorum longus
	peroneus group

Step 19. To find the soleus separate the deep muscle from the gastrocnemius on the lateral

Step 20. The extensor digitorum longus and peroneus group are located on the lateral side. The extensor digitorum longus is the first muscle lateral to the tibialis anterior.

The peroneus group is the group of muscles immediately lateral to the extensor digitorum longus.

The flexors are located immediately medial to the tibialis anterior ... If you are unable to locate the margins of the tibialis anterior ask for help, since all of the muscles on the lateral side are located by their order from the tibialis anterior.

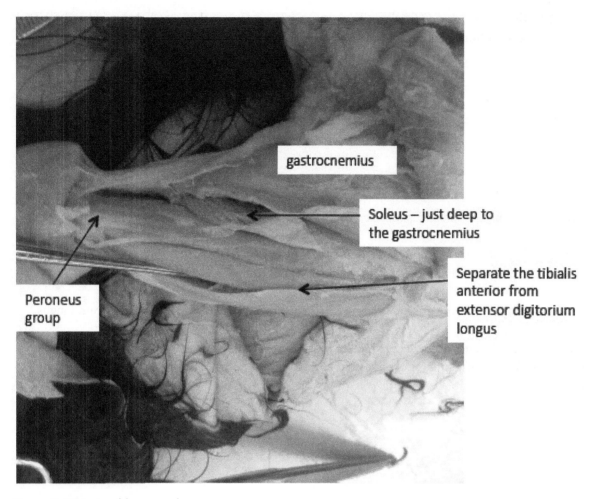

Figure 5.10 Lateral leg muscles.

WORKSHEET VI

Muscles of the Abdomen, Pectoral, Arm, Shoulder, and Back

Dissect and identify the following muscles in bold print. Know the origin, insertion, and action of muscles with an *. Be able to identify the muscles on the cat specimens and human muscle models.

DIRECTIONS FOR SKINNING THE REMAINDER OF THE CAT

Step 1. Begin by making an incision in the lower abdominal region. This incision will allow an opening to begin separating the skin from the underlying muscle tissue.

Step 2. Use your fingers to find the separation point between the skin and underlying abdominal muscles. Once you have found the separation point between the skin and muscle, continue separating the skin at this level (cutting downwards will result in cutting through and destroying muscle tissue).

Figure 6.1 Mid-abdominal incision.

Step 3. Cut along the mid-ventral line of the cat until you reach the sternum area. Then go back down to the original incision point and loosen the skin out laterally up towards the back region. Continue this pattern of loosening until the skin does not come away easily at the sternum area.

Step 4. The next lateral incision will be across from the lower sternum to the back region just under the armpit area. This cut will be difficult because of the cutaneous maximus' attachment to the sternum. The cutaneous maximus is a skin muscle that attaches the skin to the body. You will not have to identify the cutaneous maximus but it does also attach to a pectoral muscle you have to identify and to the latissimus dorsi. Because it attaches to these other muscles, it is very easy to cut these muscles and leave them attached to the folded-back skin rather than the body muscle tissue. Therefore at this point try to leave as much muscle tissue as possible on the cat body and try to leave very little on the skin.

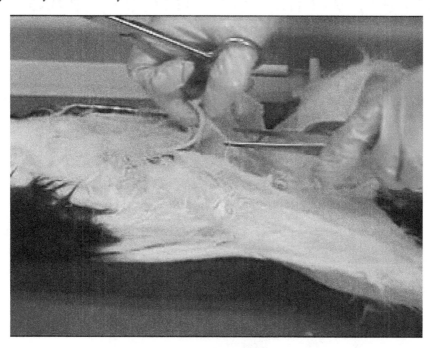

Figure 6.2 Abdominal incision into the sternum.

Step 5. Once you have made the lateral cut across the armpit, the skin over the pectoral region will come away from the muscle tissue much easier. (Be sure not to cut through the xiphihumeralis at this point.) Continue skinning up the mid-ventral line of the cat until you get to the throat region.

Step 6. As the skin is separated in the pectoral/axillary region it is possible to accidently slip under the latissiumus dorsi and separate that muscle from the other muscle tissue. At this point in skinning keep most of the muscle tissue on the cat and away from the skin. A skin muscle in this region adds to the difficulty of skinning this area. The Cutaneous maxiumus adheres the skin to the areolar connective tissue. The Cutaneous maxiumus is a thin muscle and will stay on the skin (do not try to separate it off the skin, it will be too difficult and take too long). The best way around this difficulty is to find the latissiumus dorsi and bisect it to reveal the deep muscles. This will be done after all of the skin has been separated from the muscle tissue. So at this point in the dissection try to keep the muscle tissue on the cat and the skin separated away. The best method to keep the muscle on the cat is to pull up on the skin and clip at the margin between the skin and underlying muscle.

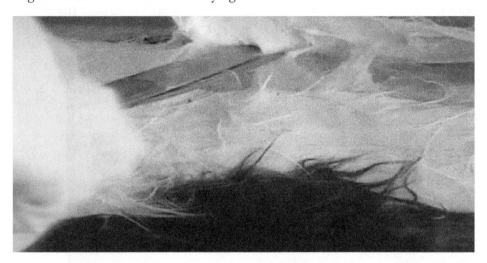

Step 7. After reaching the throat region, go back to the armpit region and cut transversely, then make a sagittal cut across the back scapular region. Continue separating the skin from the muscle tissue up to the base of the neck.

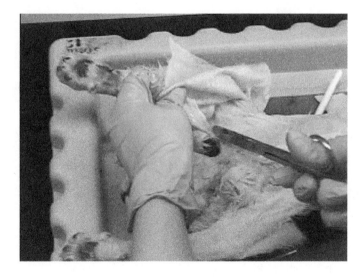

Figure 6.3 Incision pattern around sternum and forelimb.

Step 8. From the base of the neck cut transversely towards the mid-ventral line, which was loosened in a previous step.

Step 9. By completing the separation of the neck skin, the forearm area should be cleared away easily by folding the skin back and cutting the connective tissue along the muscle/connective tissue margin.

Step 10. At this time during the dissection enough has been removed to start the abdominal, back, arm and throat region.

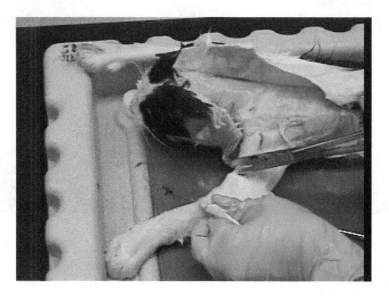

Figure 6.4 Incision pattern around back and forelimb.

Muscles of the Trunk

rectus abdominis
external abdominal oblique
internal abdominal oblique
transversus abdominis
*quadratus lumborum * only on human model
diaphragm (internal between thorax and abdomen)—
this muscle will be dissected during the dissection of the respiratory system.

*intertransversarii * only on human model
External and internal intercostals *

To view the muscles of the trunk you will have to find the rectus abdominis and the superior margin of the External oblique. The rectus abdominis is a superficial muscle with fibers running parallel with the midline (it goes from the pubis to the sternum). The best place to find the parallel fibers is at the inferior point of the sternum. Do not try to locate the rectus abdominis in the mid-portion of the abdomen, the linea alba is too thick to cut through effectively. Once you have located the rectus abdominis the superior portion of the external oblique is located just lateral. You will have to cut the oblique margin. Use the sharp tip portion of the scissors to slide under the external oblique and cut the margin along the entire middle margin. This will allow the external oblique to flap back and expose the internal oblique and transverse abdominis. The internal oblique does not extend to the midline, so you should be able to locate the margin between the two muscles without clearing to the intestines.

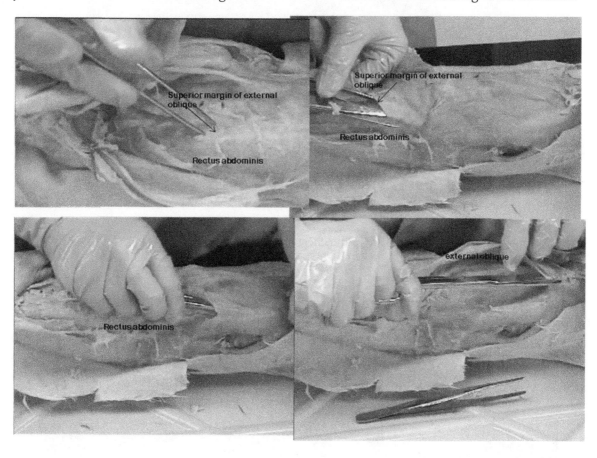

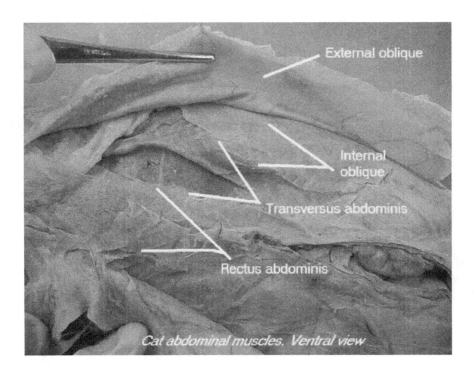

Cat abdominal muscles. Ventral view

The next group of muscles to locate are in the chest area:
* serratus anterior (ventralis)
* pectoralis major
* pectoralis minor (know the difference in the human origin and insertion)

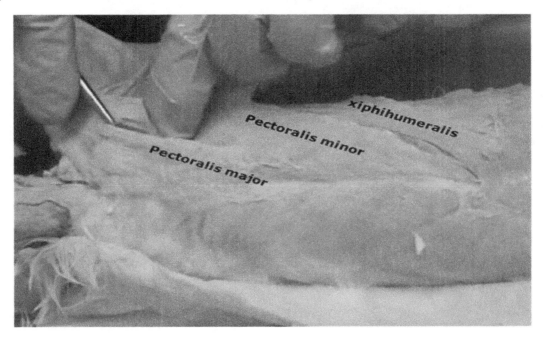

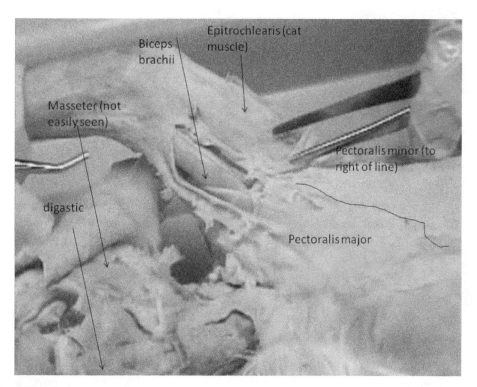

Figure 6.5 Superficial chest muscles.

Find the margin between the spinotrapezius and latissiumus dorsi.
Slide the forceps or blunt probe under the latissiumus dorsi and bisect.
This will allow the superficial latissiumus dorsi to reflect back and you can view the deep chest and shoulder muscles.

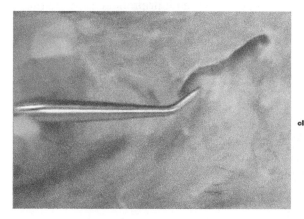
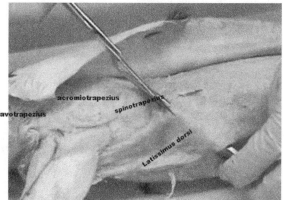

trapezius as a group
latissimus dorsi
levator scapulae
Rhomboideus (major and minor in the human). Know only Rhomboideus in the cat.

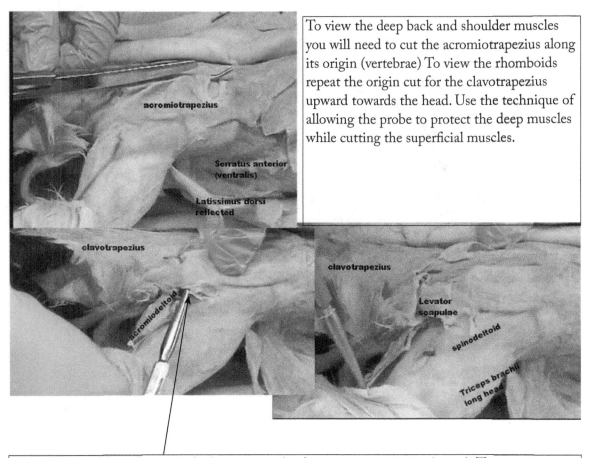

To view the deep back and shoulder muscles you will need to cut the acromiotrapezius along its origin (vertebrae) To view the rhomboids repeat the origin cut for the clavotrapezius upward towards the head. Use the technique of allowing the probe to protect the deep muscles while cutting the superficial muscles.

The harder muscle to dissect is the levator scapulae (omotransversarius in the cat). This muscle originates on the skull and inserts on the scapula. But it is deep to the clavotrapezius for a portion. Therefore you should bisect the clavotrapezius because it would damage the levator scapulae. Find the fibers that run parallel to the midline down to the scapula. Use the blunt probe to separate the transverse fibers of the trapezius from the parallel fibers of the levator scapulae. As you loosen the tissue the shape of the levator scapulae will be easier to see. The levator scapulae will be fibrously attached to the acromiodeltoid (you try to separate the two muscles but if too damage would be done it is not necessary.

The next group of muscles are on the lateral side of the cat forelimb (except for the biceps brachii which should have been exposed while viewing the superficial pectoral muscles. The coracobrachialis is difficult to locate in the cat, so we will view it on the human models.

Muscles Between the Pectoral Appendage and Vertebral Column of Thorax

* triceps brachii long, lateral, medial heads
  coracobrachialis know only on human

* biceps brachii
* brachialis
  extensors of wrist and fingers as a group
  flexors of wrist and fingers as a group

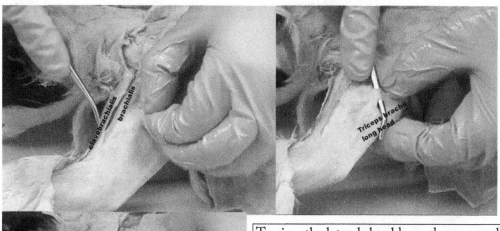

To view the lateral shoulder and arm muscles you will need to separate the margin of the clavobrachialis. There is a vein in this area but it is not a vein needed for the circulatory system so if you have to it is ok to cut the vein. Clearing the connective tissue and separating the clavobrachialis will allow the brachialis to be visible. The acromiodeltoid will also be visible.

To identify the triceps brachii medial head you will need to bisect the lateral head. The connective tissue in the area can be fibrous and may require scissors to cut. The probe will be helpful in finding the margin edges between muscle bundles.

Muscles of the Shoulder

* deltoid [as a group]
* supraspinatus
* infraspinatus
* teres major
* subscapularis

The remaining muscles of the shoulder are visible after clearing and reflecting back the acromiodeltoid and latissiumus dorsi.

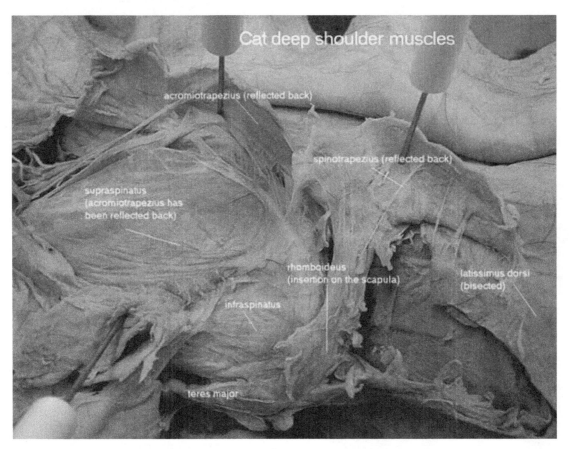

Figure 6.7 Deep lateral shoulder and back.

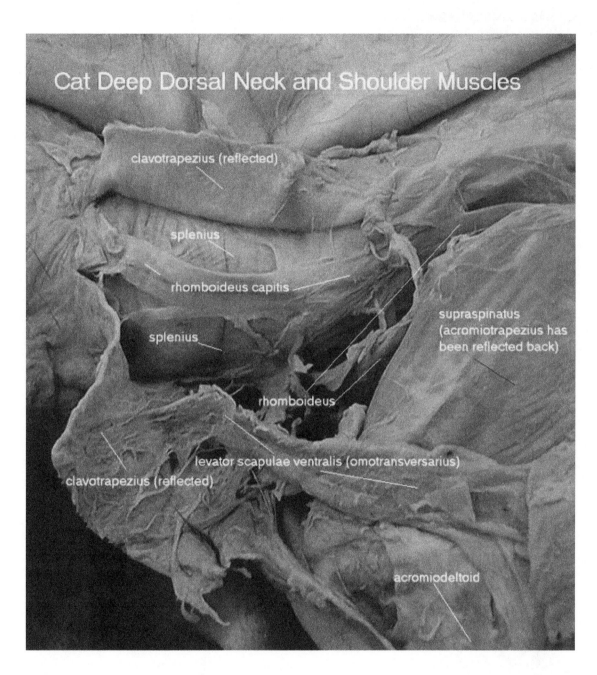

Figure 6.8 Deep shoulder and neck muscles.

Human vs. Cat muscles. Memorize the human names only.

Human	Cat
trapezius	3 parts clavotrapezius, acromiotrapezius, spinotrapezius
deltoid	3 parts clavobrachialis, acromiodeltoid, spinodeltoid
pectoralis major	pectoralis superficialis plus pectoantebrachialis
pectoralis minor	pectoralis profundus (pectoralis minor plus xiphihumeralis
sternocleidomastoid	sternomastoid plus cleidomastoid
adductor magnus plus brevis	adductor femoris

WORKSHEET VII

Muscles of the Head and Neck— Integumentary system

Dissect and identify the following muscles in bold print. Know the origin, insertion, and action of muscles with an *. Be able to identify the muscles on both cat specimens and human muscle models.

Muscles of the Head

masseter
*temporalis
digastric
orbicularus oris
(human model only)

Muscles of the Neck

sternocleidomastoid
*trapezius
splenius capitis

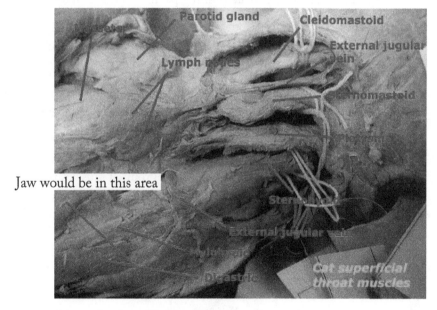

Figure 7.1 Superficial head and neck muscles.

HEAD AND NECK MUSCLE DISSECTION

Step 1. Clear the skin back from the chin up to the ear. Be careful not to clear away fat or connective tissue until the skin is away. In the cheek area the connective tissue contains glands. **If you see blue latex stop at that level and do not go deeper**.

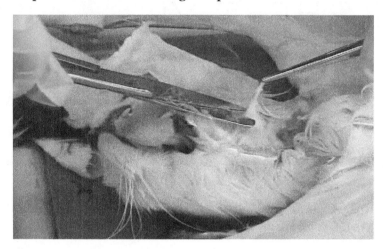

Step 2. With the forceps carefully clear away connective tissue until you see a round shaped muscle with white connective tissue fibers. This is as far as you need to go to see the **masseter** muscle.

Step 3. Find the two dentary bones (the mandible). Clear tissue away from the area with the forceps until you see muscle tissue running parallel to the dentary bone. This is the **Digastric**.

Step 4. There is a muscle that appears to span between the two digastric muscles. This muscle is at a slightly lower depth than the digastircs. It is the **mylohyoid**.

Step 5. Cut the **mylohyoid** at the margin point with the **digastric** (<u>not at the center</u>). Fold the **mylohyoid** up to expose or see the muscle lying just underneath. The fibers of this muscle will be parallel with the mid-line of the body. This muscle is the **geniohyoid**.

Step 6. There is a division between the floor of the mouth and the throat region (location of the larynx and trachea). This division is recognized by either the end of the mandibles or the transverse jugular vein (a blue latex line). The next group of muscles is caudal to this division. Locate the muscle fibers running parallel with the midline of the throat. There should be two muscles (they look like one muscle with a clear division down the middle) these are the **sternohyoids**. (Which conveniently run from the sternum to the hyoid)

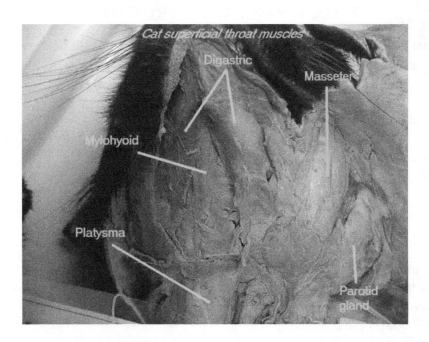

Step 7. You should see that the sternohyoids are a little deeper to two other muscles these are the **Sternomastoids**—This muscle is V-shaped (each sternomastoid muscle forms a side of the V. The fibers will not be visible because of the platysma , but the muscle is located between the blue latex line in the throat region and the exposed muscles in the center (midline) throat region. The throat is open because of the circulatory injection process.

Step 8. The **Cleidomastoid** is a difficult muscle to find but not impossible. It is located laterally to the blue line (external jugular) and medially to the clavotrapezius. The best way is to work around, with a probe, the medial margin of the **clavotrapezius** and laterally to the blue latex line (external jugular). Once you have found the margin separation the cleidomastoid is a muscle with fibers running parallel to the mid-line and it is similar in appearence to the levator scapulae (ventralis).

Step 9. Finding the **temporalis**. Cut back the skin on the dorsal head region to the level of the ear. You will see a dome shape with muscle tissue just behind the ear. That is the **temporalis**. You do not have to expose the whole muscle.

Step 10. The final muscle is the **splenius**. This muscle should already be exposed if you have bisected the **clavotrapezius**. Deep to the **clavotrapezius** is a muscle with a small thin muscle running over the top. The muscle under the small thin muscle is the **splenius** (not to be confused with the **scalenes** group on the ventral chest). The small thin muscle is the rhomboideus capitis (part of the Rhomboideus group).

Now you are finished with the Head and Neck dissection!

INTEGUMENTARY STRUCTURES

These terms are for the <u>next lab</u> and should be located on the model available in the room. Use the available photographs of skin slides to identify the layers of the skin.

Sweat gland
Sebaceous gland
blood vessels
hair follicle w/ arrector pili muscle [smooth muscle]
adipocytes
stratified squamous epithelium

Epidermis
    Stratum corneum
    Stratum lucidum (thick skin only)
    Stratum granulosum
    Stratum spinosum
    Stratum basale
Dermis
Meissner's corpuscles

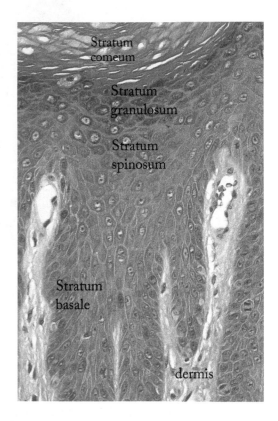

Figure 7.2 Epidermal layers.

Copyright © Nephron (CC BY-SA 3.0) at http://commons.wikimedia.org/wiki/File:Lichen_simplex_chronicus_-_very_high_mag.jpg.

Worksheet supplemental for Integumentary System:

Worksheet VII listed the structures to identify:
Be able to identify on the diagram and lab skin model the following structures:
- Sweat gland
- Sebaceous gland
- blood vessels
- hair follicle w/ arrector pili muscle [smooth muscle]
- adipocytes
- stratified squamous epithelium
- Epidermis
- Dermis
- Meissner's corpuscles [these are different from the Merkel cells in the epidermis]

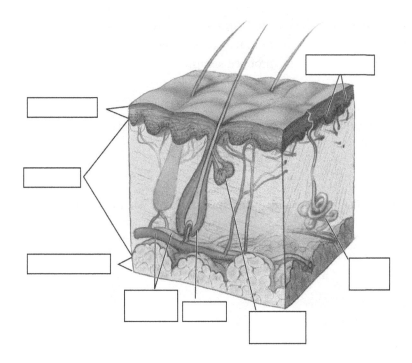

Figure 7.3 Integumentary structures to label.

Lobo, "Integumentary structures to label," http://commons.wikimedia.org/wiki/File:Anatomy_The_Skin_-_NCI_Visuals_Online.jpg. Copyright in the Public Domain.

Use this supplemental worksheet to review lecture material with the lab material requirement of identification. Each set of questions will require viewing the plates available in lab and online [the online plates will not have the unknown letter designations].

Plate I

1. Are the cells (C) in layer A alive or dead? _____

2. Are the cells (D) in layer B alive or dead? _____

3. Identify layer A _____

4. Identify layer B _____

5. What is the function of the Merkel cell found in the stratum spinosum?
    A. produce melanin   C. immune system function
    B. sensory function   D. produce keratin

Plate II

6. The skin illustrated in Plate II is which type of skin?
    A. Thick skin   C. Mesodermal skin
    B. Thin skin   D. Facial skin

7. The Dermal papilla illustrated results in the appearance of what characteristic on the epidermis?

    _____

Plate III

8. Identify layer A   _____

9. Identify layer B   _____

Plate IV

10. Identify layer A   _____

11. Identify layer B   _____

Plate V

12. Identify layer A   _____

13. Identify layer B   _____

14. Which layer of the dermis contains the blood vessels that provide oxygen and nutrients to the epidermis?

15. Which layer of the dermis would contain the illustrated sebaceous gland?

Plate VI

16. Identify muscle A.   _____

17. Which type of muscle cells compose muscle A?   _____

18. Identify structure B.   _____
19. Structure B are derived from cells from which layer?

       Layer C   or   Layer D

Plate VII

20. Identify structure C.   _____

21. Identify structure F.   _____

Plate VIII

22. Identify layer A.   _____

23. Identify layer B.   _____

24. Identify layer C.   _____

25. Identify layer D.   _____

Plate IX

26. What type of tissue is illustrated for Tissue A?
    A. Muscle tissue
    B. Connective tissue
    C. Epithelial tissue
    D. Nervous tissue

27. Identify layer B. _____

Plate X

28. Identify layer A. _____

29. Identify layer B. _____

30. Which layer contains Langerhans cells? _____

31. What is the function of the Langerhans cells? _____

MUSCLE REVIEW WORKSHEET

ACTIONS OF MUSCLES

Muscles perform their function by either isotonic contraction or isometric contraction. They are restored to their resting length upon relaxation by an antagonistic force that operates in a direction opposite to the direction of contraction. Muscles are frequently arranged in antagonistic groups such that one muscle pulls a limb in one direction, and its antagonist pulls it in the opposite direction and restores the resting length.

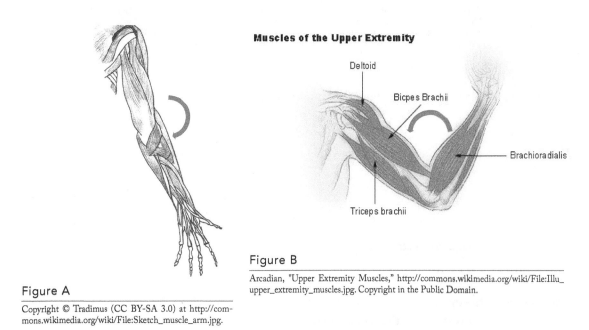

Muscles of the Upper Extremity

Deltoid

Bicpes Brachii

Brachioradialis

Triceps brachii

Figure B

Arcadian, "Upper Extremity Muscles," http://commons.wikimedia.org/wiki/File:Illu_upper_extremity_muscles.jpg. Copyright in the Public Domain.

Figure A

Copyright © Tradimus (CC BY-SA 3.0) at http://commons.wikimedia.org/wiki/File:Sketch_muscle_arm.jpg.

The diagram below illustrates how muscles are attached and how contraction results in muscular action. The muscle action illustrated in figure A is flexion / extension / adduction / abduction?

The muscle action illustrated in figure B is flexion / extension / adduction / abduction?

Draw an arrow on diagram C indicating the direction the femur would move during flexion of the thigh.

Draw an arrow on diagram C indicating the direction the tibia would move during flexion of the leg.

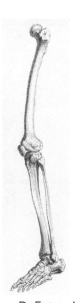

Figure C Flexion of the hindlimb

William Cheselden, Gerard Van de Gucht and Mr. Shinevoet, "Flexion of the hindlimb," http://commons.wikimedia.org/wiki/File:William_Cheselden_legs.jpg. Copyright in the Public Domain.

Figure D Extension of the hindlimb

William Cheselden, Gerard Van de Gucht and Mr. Shinevoet, "Extension of the hindlimb," http://commons.wikimedia.org/wiki/File:William_Cheselden_legs.jpg. Copyright in the Public Domain.

Use the above diagrams to understand the actions of flexion and extension of the hind limb.

If a muscle extends the thigh what bone would that muscle insert upon?

Which muscles make up the quadriceps femoris?

The muscles of the quadriceps femoris extend the leg; What bone does the quadriceps femoral tendon insert upon ? Is this the bone that is moved by the muscles of the quadriceps femoris group?

There are three muscles that together flex the leg, they all have a common origin and insertion. Identify the origin_____. ... Insertion _____

List the three muscles that flex the leg: _____

_____

_____

Synergist muscles and Fixators:

Identify two muscles responsible for stabilizing the scapula during abduction of the humerus [remember these muscles serve as fixators not synergists]

These fixator muscles should not have an attachment on which bone?
A. Scapula
C. humerus
B. Vertebrae
D. Occipital bone

Identify a muscle that serves as a synergist to the abduction of the humerus by the deltoid muscle.
A. Supraspinatus
C. Coracobrachialis
B. teres major
D. Rhomboideus major

Which two muscles flex the trunk at the hip joint?

What is the common insertion of these two muscles that flex the trunk across the hip joint?

Muscle Origins and Insertions

Action	Origin	Insertion	Muscle
Extend leg	_____	_____	**rectus femoris**
	_____	_____	**vastus medialis**
	_____	_____	**vastus lateralis**
	_____	_____	**vastus intermedius**
Extend leg and Flex hip [thigh]	_____	_____	**rectus femoris**
Flexes at the hip and flexes at knee			**Sartorius**
flex leg and extends thigh	_____	_____	**semimembranosus**
	_____	_____	**semitendinosus**
	_____	_____	**biceps femoris**
Extends thigh And lateral Rotation at hip			**Gluteus maximus**
Flexes thigh **Flexes at the hip joint**	_____	_____	**iliacus**
	_____	_____	**psoas major**
Adducts the thigh at the hip joint			**Adductor group** • longus • magnus • brevis
	_____	_____	**Gracilis**
Abducts and rotates at the hip joint			**Gluteus medius**
			Tensor fasciae latae

Moves the foot	_____	_____	
Plantar flexion			Gastrocnemius Soleus
Plantar flexion and inversion			Tibialis posterior Flexor digitorium longus Flexor hallucis longus
Plantar flexion and eversion			Peroneus longus Peroneus brevis

Dorsoflexion and inversion			**Tibialis anterior**
And extends toes			**Extensor digitorium longus** **Extensor hallucis longus**

Muscles that move the Scapula

Action	Origin	Insertion	Musclee
Adduction of the Scapula elevate, rotate, or depress scapula			**trapezius**
			rhomboideus (major/minor)
Elevates the Scapula			Levator scapulae
moves scapula down and forward protracts scapula and laterally upward {punching muscle}			**pectoralis minor** **Serratus anterior**

Muscles that move the humerus

abducts humerus			deltoid [as a group] flexes and extends humerus depending on anterior and posterior portions
			supraspinatus
Lateral rotation of humerus			Infraspinatus Teres minor
Adducts and extends humerus			Latissimus dorsi Teres major
medial rotation			Subscapularis And posterior head of deltoid
adducts humerus or flexes humerus			pectoralis major and anterior head of deltoid Coracobrachialis
extension of forearm			triceps brachii
flexes forearm			biceps brachii brachialis brachioradialis

closes jaw (elevates mandible)			**Temporalis** **Masseter**
Flexes vertebrae laterally			**intertransversarii**
lateral flexion of the lumbar spine			**quadratus lumborum**

Matching:

Match the agonist muscle in column I to its antagonist muscle in column II.

Column I   **Column II**

Column I		Column II	
Triceps brachii	_____	A.	Rectus femoris
Semimembranosus	_____	B.	Coracobrachialis
Vastus lateralis	_____	C.	Digastric
Supraspinatus	_____	D.	Biceps femoris
Temporalis	_____	E.	Biceps brachii

Levers:

There are several different types of simple machines. Sometimes you will see them classified into six different types of simple machines, while other times you may find them lumped together into various other categories based on similarities between them. The six basic simple machines are: Inclined planes, Wedges, Screws, Levers, the Wheel-and-Axle, and Pulleys.

Levers: Consist of a rigid structure rotating about a fixed point called a fulcrum. 3 Classes (see below) based on the relative positions of Fulcrum and forces. Examples: See-Saw, Crowbar, The human forearm.

All simple machines change the forces and distances involved in accomplishing work (remember that the physics definition of work involves applying a force over a distance, or Work = Force·distance; W=Fd). In other words, a machine can change how work is accomplished. Simple machines can do one or more of the following:

1. Change the location of a force;
2. Change the direction of a force;
3. Change the magnitude of a force;
4. Change the distance a force can operate over;
5. Change the speed with which a force acts.

MECHANICAL ADVANTAGE

The mechanical advantage of a simple machine is a unit-less index that tells you how much easier it is to do any amount of work with a simple machine, as opposed to doing it without the simple machine. It is a useful number in telling you things like how long a lever you need to have or how many moveable pulleys you should use to accomplish a particular task.

In dealing with simple machines we need to remember some basic terms: the <u>resistance force</u> (F_r) is just the amount of force that the machine needs to supply, and the <u>effort force</u> (F_e) is the amount of force supplied to the machine. There is a generic equation that will get you the mechanical advantage for any simple machine: You need to know (or be able to calculate) the resistance force and the effort force. Then the Mechanical Advantage (MA) is just the ratio of the resistance force to the effort force (MA = F_r/F_e – notice that the units cancel, yielding a number with no units).

For example, if we wanted to calculate the mechanical advantage of a ramp, we would need to find the force necessary to lift the mass without the ramp(= F_r), then find the force necessary to lift the mass using the ramp (= F_e); then MA = F_r/F_e.

For each simple machine, there is also a "shortcut" method for finding the mechanical advantage. Which method is most useful depends upon the particulars of the problem; generally if you are given distances, the shortcuts are faster, but if you are given forces without distances, the generic equation is faster – of course you may also be given a mix and have to find forces or distances first, and work from there.

Levers: For all levers, 1st, 2nd and 3rd class, the same rule applies:

$$\text{Mechanical advantage} = \frac{\text{Length of effort arm (distance from fulcrum to the effort force)}}{\text{Length of resistance arm (distance from fulcrum to resistance force)}}$$

Remember as well some properties of MA for the different types of levers:
1st Class levers (fulcrum between the forces): MA can be <1, = 1, or > 1.
2nd Class levers (resistance force between effort force and fulcrum): MA always > 1.
3rd Class levers (effort force between resistance force and fulcrum): MA always < 1.

How do simple machines relate to work?
- Work is the manifestation of energy
  — A force applied over a distance
  — W = F d
  — Force measured in newtons = Kg m/s^2
  — Units of Work are: Joules = kg m^2/s^2

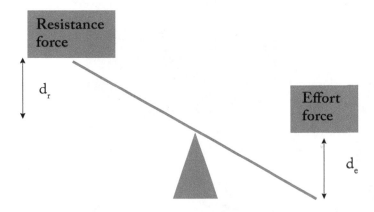

$W = F d$

$F_e D_e = F_r D_r$

If d_e or d_r is the vertical distance (movement against gravity) then how can you find horizontal movement?
- You can measure the distance of the effort and resistance arms (this should give you the same ratio as d_e:d_r) within the lever

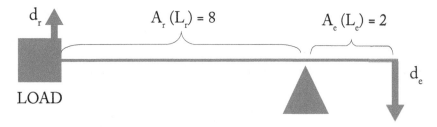

TRICEPS—FOREARM EXTENSION

The arrows indicate the direction of the force. When the triceps brachii contracts, the forearm will be moved in the arrowed direction. Additionally, any object held in the hand will exert a force in the downward arrowed direction. Ae is the effort arm, measured from the point of the effort force to the fulcrum. Ar is the resistance arm, measured from the point of the resistance force to the fulcrum. This is a different type of lever; what is different from the previous lever?

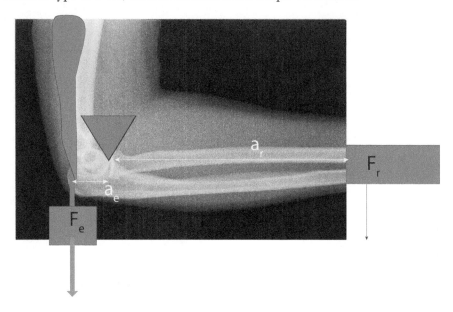

What is the mechanical advantage of a lever with a shorter a_e than a_r?

> 1 or < 1

This is a different type of lever; what is different from the previous lever?

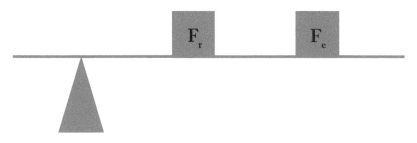

This is the last type of lever; what is different?

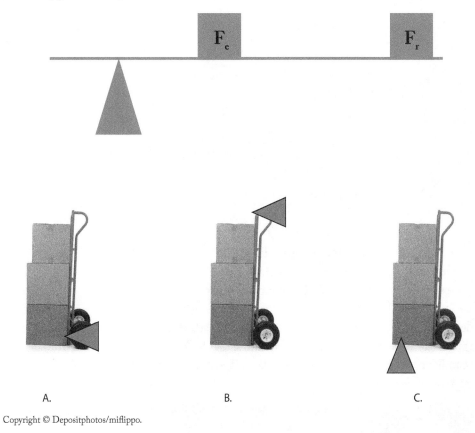

Copyright © Depositphotos/miflippo.

45. Which diagram illustrates the correct placement of the fulcrum when a utility dolly is used as a lever?

46. A pair of scissors is an example of which kind of lever system?
    A. 1st class   B. 2nd class   C. 3rd class

47. A hockey stick is an example of which kind of lever system?
    A. 1st class   B. 2nd class   C. 3rd class

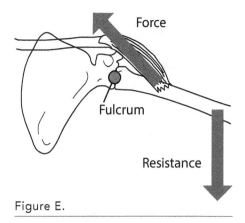

Figure E.

Identify the class of lever illustrated in figure E.
A. **First class lever**   C. **Third class lever**
B. **Second class lever**   D. **Fourth class lever**

Levers designed for strength increase the mechanical advantage of the effort force and levers designed for speed will have a low mechanical advantage. Which type of lever is illustrated in Figure E; a lever that can lift a resistance force with a smaller effort force or a lever that applies an effort force to increase the speed of the resistance arm?

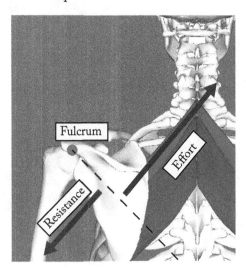

Figure F. This illustration diagrams the lever system for a deep back muscle. The arrows indicate the position of the forces relative to the fulcrum. The arrow point indicates the direction of the forces. Given these relationships, which of the following statements is true of this lever system?

Copyright © Anatomography (CC BY-SA 2.1 Japan) at https://commons.wikimedia.org/wiki/File:Rhomboid_muscles_back.png.

A.   This lever system will have *more strength than speed* for the movement of the resistance force.
B.   This lever system will have *more speed than strength* for the movement of the resistance force.
C.   This lever system will require the effort force to exceed the resistance force before the resistance force will be moved.
D.   The resistance force will move a greater distance than the effort force required to move it.

Figure G.

Mikael Haggstrom, "Muscles posterior," https://commons.wikimedia.org/wiki/File:Muscles_posterior.png. Copyright in the Public Domain.

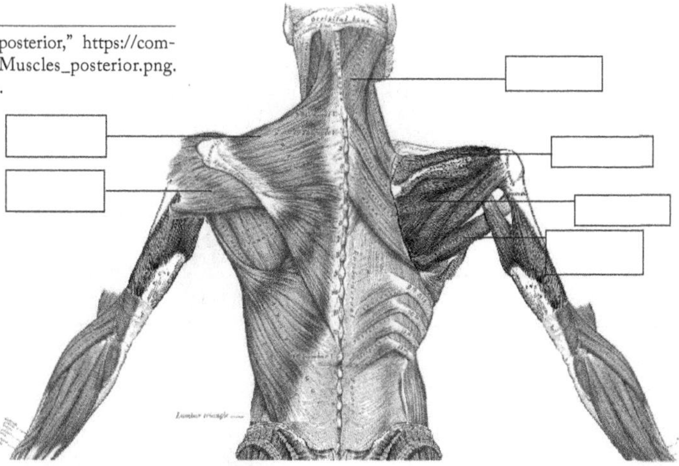

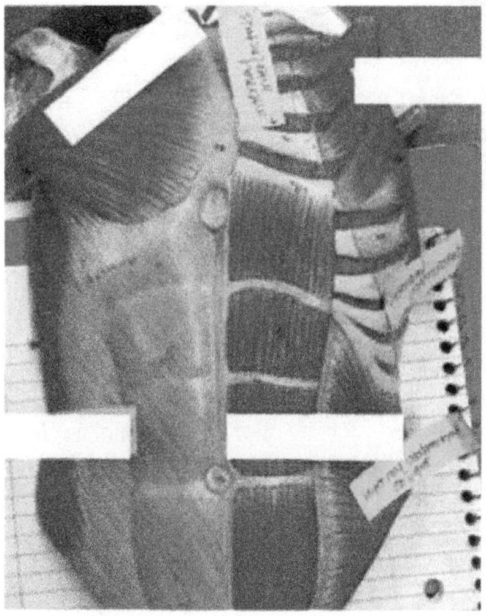

Figure H.

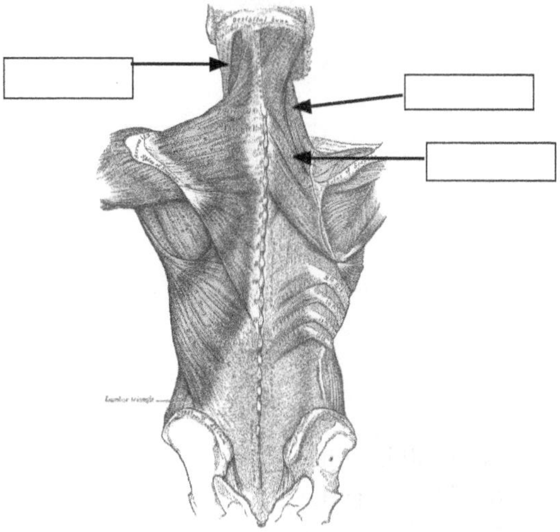

Figure I.

Henry Gray, "Gray409," https://commons.wikimedia.org/wiki/File:Gray409.png. Copyright in the Public Domain.

Prime movers of scapula abduction	1	
	2	
Prime movers of scapula adduction	1	
	2	
Prime movers for arm abduction/flexion	1	
	2	
Prime movers for arm adduction/extension	1	
	2	
	3	
	4	

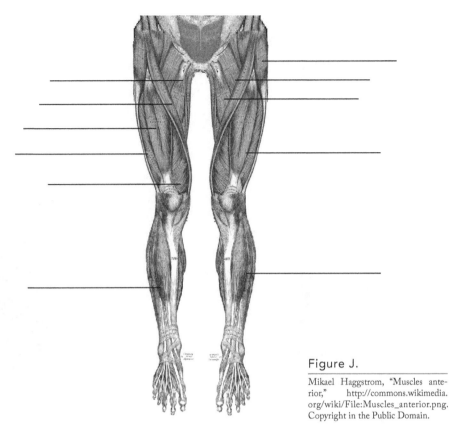

Figure J.

Mikael Haggstrom, "Muscles anterior," http://commons.wikimedia.org/wiki/File:Muscles_anterior.png. Copyright in the Public Domain.

Note: all leg muscles are not labeled.

6. Where is the gracilis?
7. Is the vasturs lateralis under the tensor fascia latae?
8. Where is the soleus?

For further studying you should photocopy a picture from your book and label the muscles.

MUSCLE IMAGES LABELING

Identify all muscles from your check list on the following images.

Artery and vein

Figure 0 – Medial thigh muscles

<u>Muscles in view, deep to the Gracilis</u>

Adductor longus, Adductor femoris, Semimembranosus, and Semitendinosus

<u>Muscles in view, deep to the Sartorius</u>

Vastus medialis, Rectus femoris, and (small portion of) vastus lateralis. There is a small portion of the iliopsoas visible.

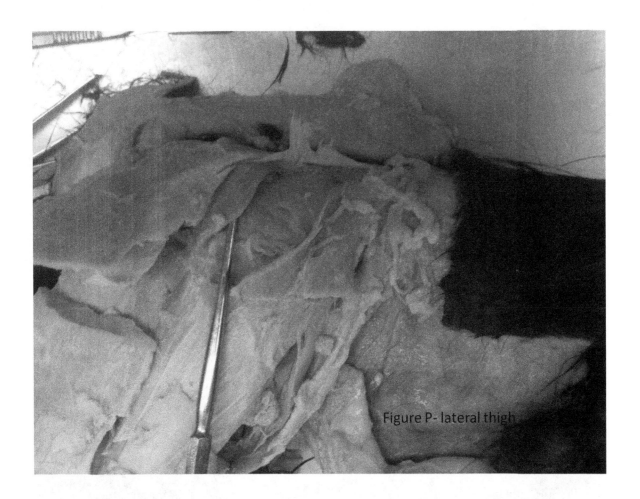

Figure P- lateral thigh

<u>Muscles in view, lateral thigh</u>

biceps femoris, gluteus medius, tensor fasciae latae, and vastus lateralis (under the iliotibial tract).

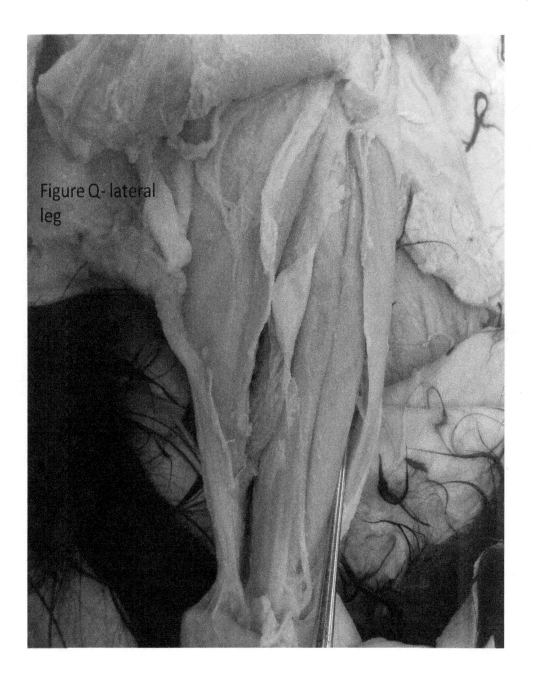

Figure Q- lateral leg

<u>Muscles in view, lateral leg</u>

gastrocnemius, soleus, fibularis (peroneus) group, extensor digitorium longus, and tibialis anterior.

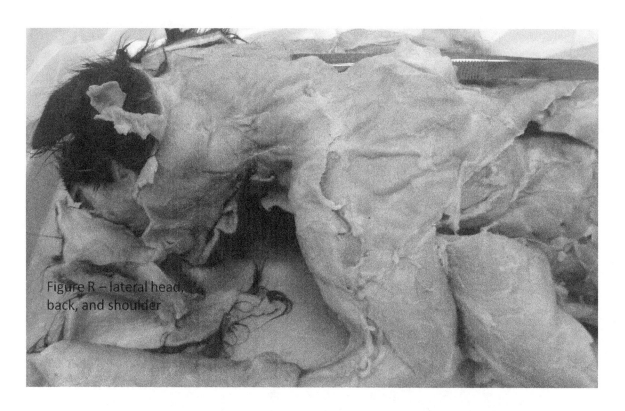

Figure R – lateral head, back, and shoulder

Muscles in view lateral head, back and shoulder

trapezius, masseter, levator scapulae, deltoid, brachialis, triceps brachii (lateral head and long head), latissiumus dorsi, teres major, and serratus anterior.

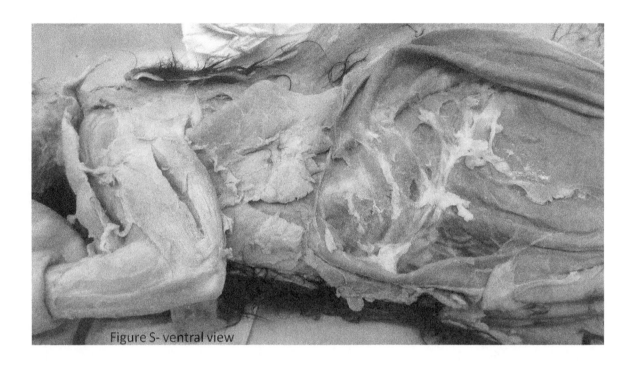
Figure S- ventral view

Muscles in view, ventral

extensors of the wrist and fingers, external intercostals, rectus abdominis, external abdominal oblique, internal abdominal oblique and transverse abdominis.

WORKSHEET VIII

Respiration, Digestive, Urinary, Heart,and Circulatory Lists

CAT VISCERA

Note: In addition to their identity, be able to state the general function of the structures indicated below. In the case of blood vessels, know the structures served and the blood flow route. Dissection instructions and images follow the check list on the next page.

<u>Locate the following on the models</u>

External nares, nasal cavity, internal nares, oral opening, tongue, palate, oral cavity, pharynx, Eustachian tubes (locate on human skull), palatine tonsils (visible in the cat)

Digestive system
- stomach
- pyloric sphincter
- small intestine
  - duodenum
  - jejunum
  - ileum

- large intestine
  - cecum
  - colon
  - rectum
- liver
- gall bladder
- common bile duct
- pancreas (head, body)
- esophageal opening
- esophagus
- parotid gland
- submandibular gland
- sublingual gland
- epiglottis

Lymphatic system
- spleen
- thymus gland
- lymph nodes
- Respiratory system
  - trachea
    - tracheal cartilages
  - bronchus
  - lung
  - diaphragm
    - larynx
    - vocal cords
      - thyroid cartilage
      - cricoid cartilage

Pleura and pleural cavities

Accessory muscles
- external intercostals
- internal intercostals
- diaphragm

Peritoneal mesenteries
- lesser omentum
- greater omentum
- mesentery

Urinary system
- kidney:
- renal cortex
  - renal medulla
  - pyramids
- urinary bladder
- urethra
- ureter

<u>Know the heart structures and their functions</u>

a=artery, aa=arteries, v=vein, vv=veins, n=nerve

Abdominal vessels
abdominal aorta
Visceral branches: (unpaired)

celiac trunk
   hepatic a
    left gastric a
    splenic a
   superior mesenteric a

Thoracic & head vessels
aortic arch >> thoracic aorta
Branches:
brachiocephalic trunk
   rt. and lt. common carotid aa
   rt. subclavian a
internal mammary aa
left subclavian a
axillary aa
brachial aa

HEART
rt. & left atria
  auricles
rt. & left ventricles

pulmonary trunk
aorta
ligamentum arteriosum
  (ductus arteriosus in fetus)
pulmonary aa
pulmonary vv
musculi pectinati
coronary sinus opening
semilunar valve
  (aortic and pulmonary)
fossa ovalis
  (foramen ovale in fetus)
chordae tendinae

Systemic branches: (paired)
abdominal aorta
renal aa
internal iliac aa
external iliac aa
femoral aa
spermatic aa
ovarian aa

superior vena cava
tributaries:
azygos v
brachiocephalic vv
internal mammary vv

subclavian vv
axillary vv
brachial vv

papillary muscles
trabeculae carnae
tricuspid valve
bicuspid valve

Systemic tributaries
inferior vena cava
renal vv
ovarian vv
spermatic vv
internal iliac vv
external iliac vv
common iliac vv

external jugular vv
anterior facial vv
posterior facial vv
transverse jugular v

Portal veins
superior mesenteric v
gastrosplenic v
hepatic portal v

Differences in the circulatory system of humans and cats	
Cat	**Human**
cranial mesenteric vein	superior mesenteric vein.
gastrosplenic vein	splenic vein
internal spermatic a. (male)	testicular a.

THORACIC CAVITY: CIRCULATORY AND RESPIRATORY SYSTEM DISSECTION

Step 1. Make two cuts lateral to the sternum. Start at the diaphragm and cut towards the neck region. As you near the last cranial rib on the left side, be cautious about the **brachial plexus** (These are the nerves and vessels running through the armpit). If you cut at the ventral portion of the rib, you should not go too deep and avoid cutting the underlying vessels.

Step 2. Cut transversely across the top of the chest region. This will separate the sternum from the thoracic cavity. Once you have removed the sternum, you will see some brown tissue overlying the tracheal tube and the arteries coming out of the heart. This tissue is the **thymus gland.** This gland is part of the lymphatic system. It is larger in younger individuals and is reduced in size in older individuals. In younger individuals the thymus serves as an important storage and producer of lymphocytes (which are part of the immune system). The **mediastinal septum** might be visible between the two pleural cavities but it is difficult to see and you should be familiar with the description of the structure. The **parietal and visceral pleura and pericardium** are similar to those in the peritoneum. The parietal covers a cavity: either the lung cavity or heart cavity. The visceral covers an organ: either the lungs or heart.

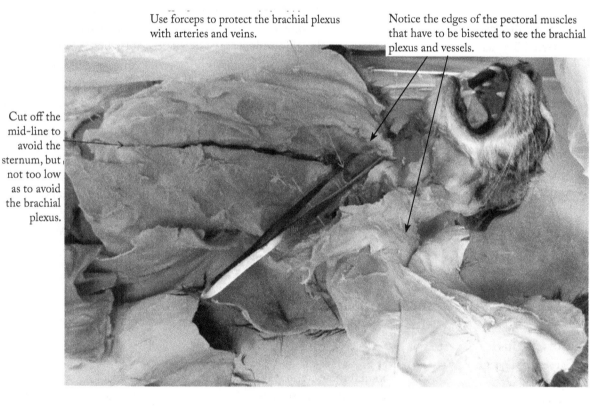

Use forceps to protect the brachial plexus with arteries and veins.

Notice the edges of the pectoral muscles that have to be bisected to see the brachial plexus and vessels.

Cut off the mid-line to avoid the sternum, but not too low as to avoid the brachial plexus.

Figure 8.1 Chest incisions for removal of sternum to view organs and blood vessels.

Step 3. Next cut the muscle tissue away from the underlying structures. Do this carefully and in layers until you begin to see blue latex. At that point you will have exposed the veins in the thoracic system.

Step 4. Before you begin to expose the vessels finish clearing muscle tissue away from the **trachea.** Remove only the top muscle tissue, the lower tissue will be removed later when you locate the vessels and nerves.

Step 5. Cutting the mental symphysis (lower jaw). Use the cutting shears and begin at the middle point of the lower jaw (in the middle of the incisor teeth). Cut the lower jaw in half to the point where the two digastric muscles meet.

Step 6. Now cut with scissors along edge of each lower dentary bone. This will separate the connections of the tongue. You should continue cutting the side muscles to the point of the caudal portion of the digastric muscles.

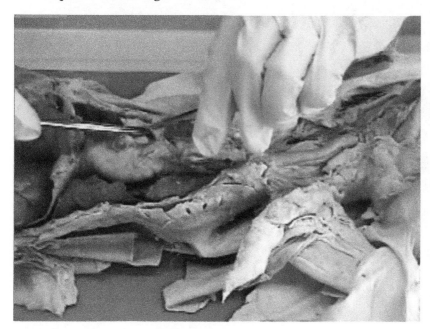

The muscle tissue will need to be cut, but the loose connective tissue can be cleared with the blunt forceps. Do not use scissors to remove the tissue from the heart and superior veins.

Step 7. Pull the tongue back to expose the **epiglottis**. This is a small protruding flap just before the **glottis** (the opening to the respiratory system).

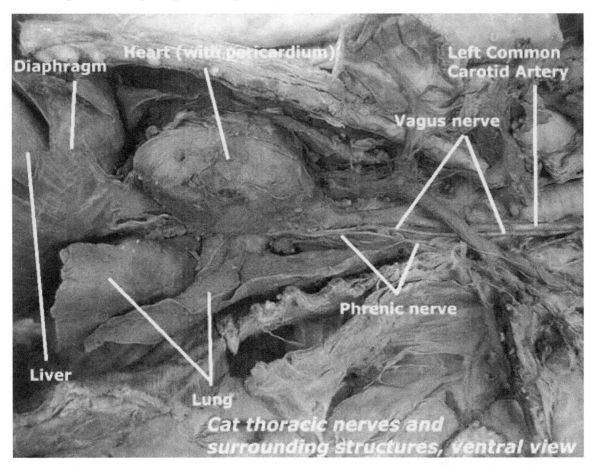

Figure 8.2 Thoracic and throat structures once the connective tissue is cleared.

Step 8. Confirm the location of the **glottis** by pushing a probe through the opening. The probe will be visible in the tracheal tube. Notice that the **esophagus** is located dorsal to the trachea.

Step 9. Notice the bulbous, round structure in the tracheal tube; this is the **larynx.** Pull the muscle tissue away from the underlying cartilage. Carefully use scissors if the muscle tissue does not remove easily. Be careful not to remove tissue from the lateral sides of the trachea. Just caudal to the **larynx** located lateral to the trachea is the **thyroid gland** (this gland functions in regulation of the metabolic rate and serves to regulate growth in younger individuals).

Step 10. Now that the muscle tissue is removed, you will see that the larynx is divided into two parts; the cranial part is the **thyroid cartilage** and the caudal part is the **cricoid cartilage.**

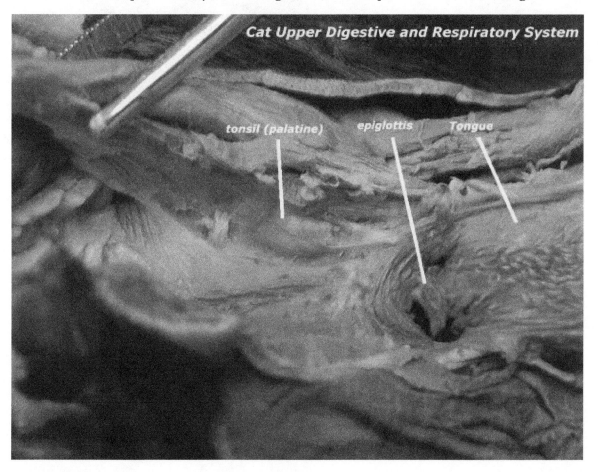

Figure 8.3 Cutting the mental symphysis for respiratory and lymphatic structures.

Step 11. Following the trachea down to the lung it will separate into two tubes. At this point are the **bronchi.** The **lungs** are the large, brown, lobed structures around the heart.

Step 12. To view the structures of the heart you have to remove the **pericardium**. Make a small incision in the top of the heart. This is the outer **pericardium**. Cut downward and peel the pericardium back.

Step 13. You should be able to identify the **right and left auricles** (the brown soft flaps located at the cranial lateral portions of the heart), the **right and left ventricles** (you do not have to cut open the heart to identify these—just know the area they are located in), the **pulmonary trunk** (this is the tube exiting the cranial part of the heart and it will not be injected with latex). The aorta will be injected and can be distinguished by the **brachiocephalic trunk** and **left subclavian** exiting the aorta. The **anterior interventricular sulcus** may not be visible because of preservative material and it will be more visible on the pig heart.

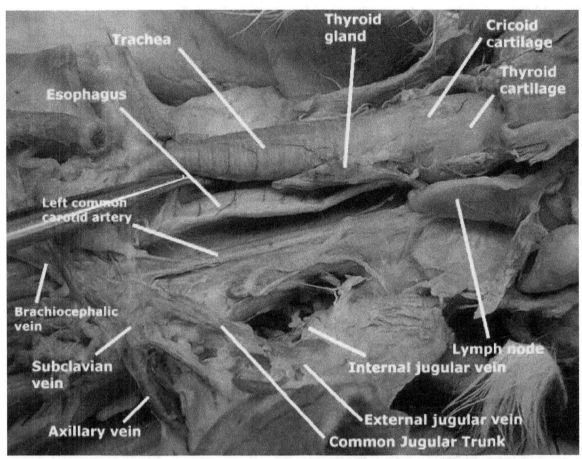

Figure 8.4 Throat structures for respiratory, digestive, and circulatory systems.

Step 14. Do not put your cat away, since you will be looking for vessels at the end of the dissection. The next set of structures are best illustrated in the pig heart.

Step 15. Use a pig or sheep heart cut in half to see the following structures:

Anterior interventricular sulcus—this is the indentation between the two ventricles on the outside to the heart.

Coronary vessels—These are the lines leading off the sulcus. Keep in mind these vessels are not injected, so they appear white.

Right and left auricles—These are the flaps off the lateral sides of the atria.

Right and left atria—These are the smaller top (superior chambers). Determining right from left is easier once you have identified the right and left ventricles.

Right and left Ventricles—These are the larger inferior chambers. The right ventricle is smaller than the left ventricle. The left ventricle is large and the muscle walls are thicker (because the need to pump blood out the aortic arch to the rest of the body). The left ventricle also contains the apex or tip of the heart. By this I mean if you look at the tip of the heart it will be the lower portion of the left ventricle. The right ventricle is oriented higher up on the heart and the tip of the chamber is in the lateral side of the heart.

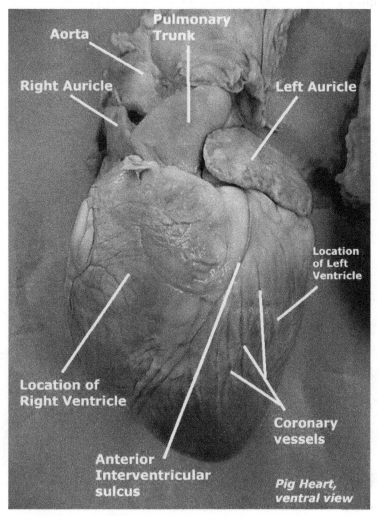

Figure 8.5 Anterior view of the external heart features.

It is important to identify the right and left side of the heart because the rest of the structures are identified by which side of the heart they are located on.

Aorta This vessel has probably been cut off the pig heart and is best viewed in the cat or on the heart model in the lab. This vessel receives blood from the left ventricle after the blood has passed through the **semilunar valve of the aorta**. The semilunar valve can be seen in the pig heart. Use a probe to follow out the left ventricle. At the top of the ventricle will be two half-moon white flaps; these are the aortic semilunar valves.

The pulmonary trunk (which branches into the pulmonary arteries) can be seen as the vessel looping over and out of the superior portion between the two atria on the ventral side of the heart. (The side you will see in the cat, because as the cat is on its back, you are viewing the ventral side of the heart).

Blood exits the right ventricle to enter the pulmonary trunk after passing the **semilunar valve** of the pulmonary trunk.

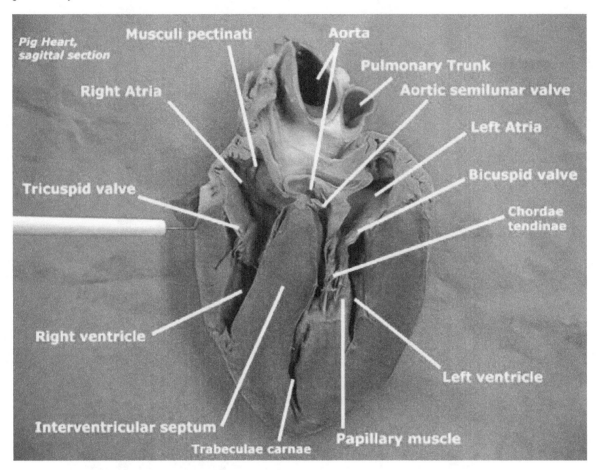

Figure 8.6 Internal anterior view of the posterior heart half.

The **inferior and superior vena cava** will be cut off the pig heart, but the holes left behind can be seen on the dorsal side of the right atria. These are best viewed on the heart model.

The **tricuspid valve** is the flap anchored by the **chordae tendineae** between the right atria and the right ventricle. The chordae tendineae are attached to the heart wall by the papillary muscles. These can be seen as protrusions at the end of the chordae tendineae.

The **bicuspid valve** is the flap between the left atria and the left ventricle.

The **fossa ovalis** is a depression in the side of the right atria. It is what is left of the foramen ovale, which is the hole between the right and left atria of the developing fetus. This hole allows the blood to bypass the right ventricle and the lungs (since the lungs are not the site of gas exchange for a fetus). This structure is difficult to see in the pig heart and is best illustrated by the heart model.

The **coronary sinus** is a hole located in the right atria wall next to the opening of the inferior vena cava. This hole is difficult to see and can be seen in the heart model. The coronary sinus is how blood that supplied oxygen to the heart is returned to the body's circulation to be re-oxygenated.

The **musculi pectinati** are the muscular ridges in the walls of the atria.

The **trabeculae carnae** are the muscular ridges in the walls of the ventricles.

THORACIC BLOOD VESSEL DISSECTION

Use your cat to find the following thoracic blood vessels.

Step 1. Start at the heart and the aortic arch. There are two light or red cords exiting the heart. These are the **brachiocephalic artery** and the **left subclavian artery**. The brachiocephalic artery is the first one on the right side of the cat.

Step 2. First, follow up the brachiocephalic artery. This artery continues up towards the head. It will branch off into three vessels. (This branch, however, is concealed by the huge blue brachiocephalic veins that branch into the superior vena cava. To see the branching off of the brachiocephalic artery, carefully clear the tissue away from the vessels and look under the brachiocephalic veins.) The first two branches from the brachiocephalic are the **right and left common carotids**. These are the two red vessels running parallel and lateral to the trachea. The third branch is the **right subclavian artery**, which delivers blood to the arm arteries. (**Notice the right and left subclavian arteries are not symmetrical. The left subclavian branches directly from the aortic arch, where the right subclavian does not branch off until after the brachiocephalic.**)

Step 3. From the right subclavian out to the brachial is symmetrical to the left side of the body. Therefore do not bother dissecting out the right side of the cat's arm, since we will see the same vessels on the left side.

Step 4. Return to the aortic arch and follow the **left subclavian** up to the left arm of the cat. Now you will have to clear away tissue from the vessels. This should be done carefully so as not to break any of the vessels. Also do not cut any white thread-like or string-like structures. These are the nerves you will learn next week. At this point it is useful to note how blood vessels are identified. When a vessel branches, the resulting vessels will have different names even if one of the branches appears to be a continuation of the previous vessel. This is the case with the left subclavian artery. The left subclavian artery at the point of exiting the rib cage region and entering the **armpit region has some vessels that branch off (none of which you have to know). The vessel that continues out to the arm is no longer the left subclavian but is termed the axillary artery. At the region of the scapula and beginning of the humerus, there are more arteries that branch off and the continuing vessel is termed the brachial artery.**

Step 5. Now return to the heart and we will follow the veins. Note the following descriptions will be to identify the vessels and do not suggest the flow of blood, since the blood in the veins will be flowing towards the heart, not away. Follow the **superior vena cava** up to the first major split. The short branches off the superior vena cava are the **brachiocephalic veins**.

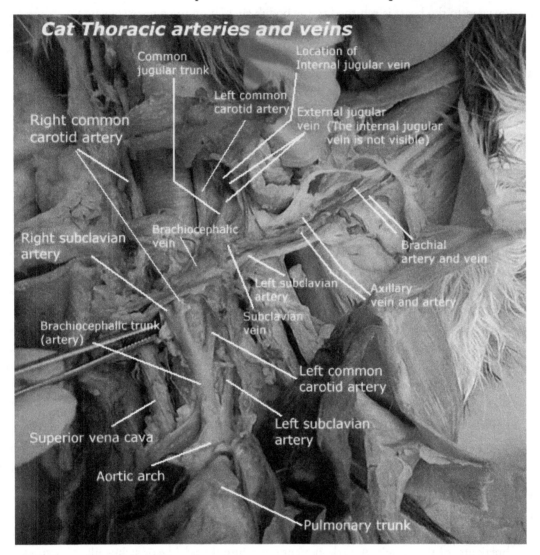

Cat Thoracic arteries and veins

Common jugular trunk

Location of Internal jugular vein

Left common carotid artery

External jugular vein (The internal jugular vein is not visible)

Right common carotid artery

Right subclavian artery

Brachiocephalic vein

Brachial artery and vein

Left subclavian artery

Axillary vein and artery

Subclavian vein

Brachiocephalic trunk (artery)

Left common carotid artery

Superior vena cava

Left subclavian artery

Aortic arch

Pulmonary trunk

Figure 8.7 Thoracic and throat vessels.

Step 6. From this point follow up on the left side (as the right side is symmetrical). The **brachiocephalic vein** next will have a branch off to the left leading up the throat region (but not as lateral as the *external jugular vein* that is very visible in the upper neck region), which is the smaller vessel of the **internal jugular vein**. *(At this point I will mention that the internal and external jugular veins do not empty directly into the brachiocephalic but empty into the common jugular trunk, which then leads into the brachiocephalic vein. But you do not have the common jugular trunk on your list, so all you have to identify is the internal and external jugular.)* The vein that continues after the internal and external jugular is no longer the brachiocephalic but is now the **subclavian vein**.

This vein very short. Once you see another vein branching off the subclavian vein, the vessel that continues out is the **axillary vein** (this is somewhere at the line of the rib cage and armpit). Then at the region of the scapula, the vein that drains the arm is the **brachial vein**.

Step 7. The last two vessels to find are located in the head region near the lymph nodes and submandibular gland. At the superior end to the external jugular, there is a branch into two vessels. The branch that directs towards the mouth is the **anterior facial vein** and the branch that is directed towards the parotid gland is the **posterior facial vein**.

CAT AND SHEEP HEARTS AS AN EXAMPLE FOR THE HUMAN HEART

Structures

Cranial and caudal vena cava drain into the right atrium. The tricuspid valve is between the right atrium and right ventricle, which prevents the backflow of blood when the right atrium pumps blood into the right ventricle. The deoxygenated blood then flows out of the ventricle into the pulmonary trunk. At the entrance of the pulmonary trunk is the pulmonary semilunar valve, which prevents backflow of blood from the lungs into the right ventricle. The blood from the lungs is now oxygenated and flows back to the heart via the pulmonary veins. The pulmonary veins dump blood into the left atrium (so the pattern of blood from the body enters the heart via the atria and pumps blood out of the heart via the ventricles is maintained in all of the vertebrates). Once the oxygenated blood is in the left atrium, it flows to the left ventricle past the bicuspid valve (which prevents backflow into the atrium). The blood is then pumped into the aortic arch past the aortic semilunar valve (which prevents backflow into the left ventricle).

Ways to learn and distinguish the vessels of the cat

The aortic arch has two main branches feeding blood to the head and arm region: The first (and larger of the two) is the brachiocephalic; the second is the smaller left subclavian.

The brachiocephalic has three branches: rt. subclavian, and rt. and lt. common carotids.

The left subclavian leads into the vessels of the arm. The ones you have to know appear continuous with the left subclavian, so the best way to learn them is by location relative to other parts of the cat's body. When the left subclavian exits the rib cage and enters the armpit region, it is the axillary artery. The axillary artery becomes the brachial artery once the vessel is past the scapula and running along the arm area (remember brachial refers to the arm region and branchial refers to the gill region).

The veins are in reverse order (because the blood flow is back to the heart). The regions are still the same for the veins: The brachial vein is in the arm and the axillary vein is in the armpit. The subclavian vein is shorter than the subclavian artery and dumps blood into the two brachiocephalic veins.

The head region of the cat is drained by the transverse facialis vein into the external jugular. The "jugulars" (both internal and external and common jugular trunk) drain blood into the brachiocephalic vein. So the two brachiocephalic veins drain deoxygenated blood from the head and arm into the cranial (superior) vena cava.

ABDOMINAL BLOOD VESSEL AND THORACIC/ ABDOMINAL NERVE DISSECTION

Step 1. Directions on how to clear away connective tissue from around the blood vessels. The peritoneal cavity has already been opened. This dissection will require you to clear away connective tissue from around the blood vessels. To do this you will primarily use blunt forceps. Use the forceps in a raking motion and pull with medium force. The larger blood vessels will not break against medium pressure, but the connective tissue and fat will give against the pressure.

Step 2. Finding the celiac and superior mesenteric arteries. The first sets of vessels are located directly under the cardiac portion of the stomach. Gently lift the stomach up and towards the right side of the cat. This will expose the back of the peritoneal cavity near the diaphragm. There are some layers of fat and connective tissue covering the back wall of the peritoneal cavity. Use the blunt forceps in a raking motion to expose the red injected arteries. You may not find the abdominal **aorta** but you should see two vessels branching off and going into the stomach and intestinal region. There are strong spider-web-type strings lying over the two vessels. These are the celiac and superior mesenteric ganglion. It is ok to remove them, since they block your view of the **celiac** and **superior mesenteric arteries**. Be careful not to cut or remove the blood vessels while trying to remove the ganglion.

Step 3. Finding the celiac artery and its branches. The first branching vessel below the diaphragm from the dorsal aorta is the **celiac artery**. Use the forceps to clear away connective tissue and expose the three branches off the celiac trunk. The first branch is difficult to see because is continues deep under the stomach and delivers blood to the liver. This first branch is the **hepatic artery**. The second branch off the celiac is smaller but goes directly to the stomach, and is the **left gastric** (no, there is no right gastric, but you should include the "left" part when answering questions on the exam). The third branch off the celiac is the **splenic artery**. This vessel will appear as a continuation of the celiac artery. (Note these three vessels form a three-pronged fork from the celiac. The hepatic is difficult to see so make sure your celiac artery has three prongs and not two.)

Step 4. Finding the first artery that branches off the superior mesenteric artery. The second branch off the abdominal aorta in this region is the **superior mesenteric artery**. The branches from this artery are not as close as those from the celiac artery, so be prepared to go looking for the following arteries. The first branch off the superior mesenteric artery is the **posterior pancreaticoduodenal artery**. This artery can be found in the connective tissue and pancreas area of the lower duodenum. You have to use the forceps in a raking motion to expose the vessel and then expose it back to where it branches off the superior mesenteric.

Step 5. Finding the second branch off the superior mesenteric artery. The second branch off the superior mesenteric artery is the **ileocolic artery**. This artery can be found by raking with the forceps the connective tissue near the cecum in the direction of the superior mesenteric artery. Clearing away this connective tissue should expose the **ileocolic artery** branching off the superior mesenteric artery and delivering blood to the ileocecal junction (the area between the ileum and colon).

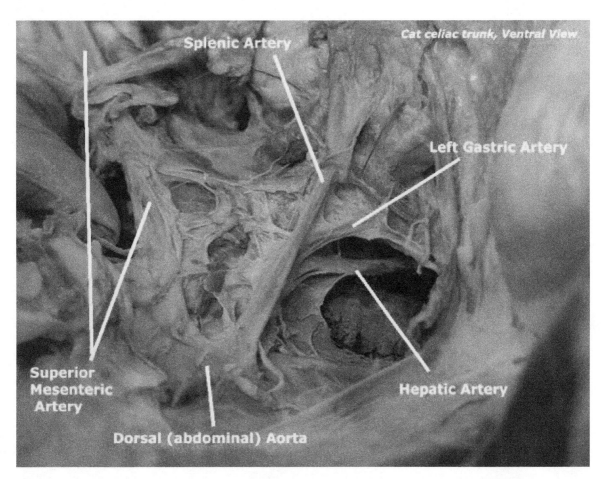

Figure 8.8 Branches off the abdominal aorta, the celiac trunk, and superior mesenteric arteries.

Step 6. Finding the third branch off the superior mesenteric artery. The third branch off the superior mesenteric is a set of arteries termed the **intestinal arteries.** These are several arteries that deliver blood to the small intestine in several areas. There are many of them and they are best visible in the clear connective tissue between curves of the small intestine.

Step 7. Finding the branches off the abdominal aorta and tributaries of the inferior vena cava

A. The celiac and the superior mesenteric branch off the abdominal aorta. The next artery to directly branch off the abdominal aorta is the **adrenolumbar artery.** This artery is very small and difficult to find. The **adrenolumbar vein** is more visible and will be the one usually tagged on an exam. This vein can be found lying over the ventral side of the adrenal gland. Note this vein drains blood directly into the inferior vena cava.

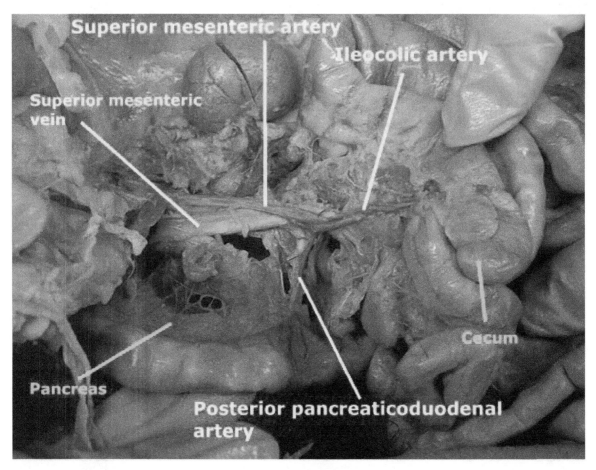

Figure 8.9 Branches off the superior mesenteric artery.

B. The **renal artery** is the next artery to branch off the aorta. This artery may be hard to see since the **renal vein** is larger and covers the artery. Both of these vessels can be found by starting from the kidney and following back towards the aorta and inferior vena cava.

C. The **genital arteries and veins** are sometimes difficult to find because they are thin and perhaps not injected. The best way to find them is to start at the gonads (either the ovary or testis) and follow the vessel back to the aorta or inferior vena cava.

D. The **inferior mesenteric artery** can be found at the lower portion of the abdominal aorta. This artery delivers blood to the large intestine. It is located between the **renal arteries and iliolumbar arteries**.

E. The **iliolumbar arteries and veins** are located in the lumbar area of the dorsal wall of the peritoneal cavity. They run together and the artery is often hard to see because the vein might run over the top of the artery.

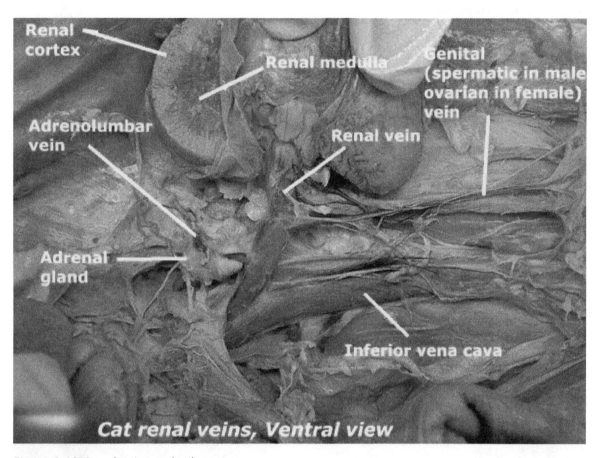

Figure 8.10 Renal veins and tributaries.

F. At the end of the abdominal aorta is a fork of two vessels. There is also at this point another two vessels that branch off. Both of these forks are difficult to see until you clear away the connective tissue away from the aorta. The urinary bladder and connective tissue may also be in the way. The large intestine and rectum may also have to be pushed to the side to find the end of the abdominal aorta. The outer two branches continue down into the inner thigh region. In the first part of this section are the **external iliac arteries.** The two inner branches are the **hypogastric arteries** (internal iliac arteries).

G. The **external iliac arteries** continue and at the point where they exit the peritoneal cavity and inter the inner thigh is the **femoral artery.** The **femoral vein** is located next to the artery in this region.

H. The **external iliac veins** do not directly drain blood into the inferior vena cava. The two vessels that drain blood directly into the inferior vena cava are the **common iliac veins. The internal and external iliac veins** can be seen in the lower pelvic region. The hypogastric or internal iliac veins are difficult to see and require the connective tissue and pelvic region to be cleared of fat and connective tissue.

I. Do not forget to identify the **inferior vena cava.** This is the major vein that returns blood to the heart, via the right atria.

Step 8. The hepatic portal system

There are three veins left to locate. These three are part of the hepatic portal system and were injected with yellow latex. The reason these vessels are injected separately from the other arteries and veins is that the hepatic portal veins drain blood from the stomach and intestines to the liver. (Note this is different from other veins that drain blood from organs to the inferior vena cava, which returns blood to the heart.)

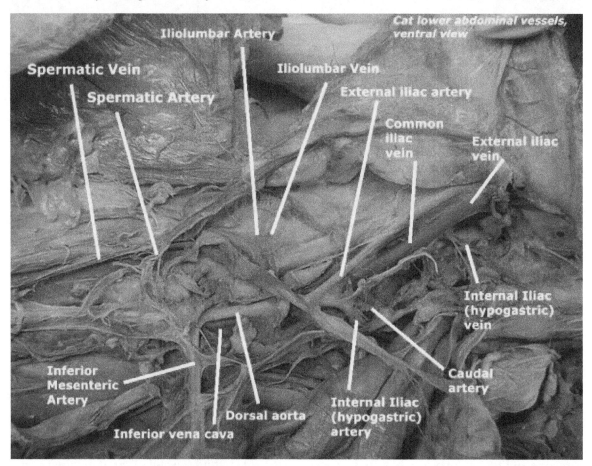

Figure 8.11 Inferior circulatory vessels.

J. The **superior mesenteric vein** can be found on the right side of the cat's peritoneal cavity. If you can find the superior mesenteric artery, you will see a yellow vessel coming out of the intestines. The yellow vessel coming out of the intestinal region is the superior mesenteric vein.

K. The **hepatic portal vein** is the yellow vessel located in the connective tissue bundle that also contains the common bile duct. To find the **gastrosplenic vein** you will need to find the hepatic portal and follow it back to the **superior mesenteric vein**. The superior mesenteric vein and the hepatic portal vein will appear continuous. At the point where the gastrosplenic vein drains blood into the hepatic portal is also the location where the superior mesenteric vein becomes the hepatic portal vein.

DIGESTIVE SYSTEM DISSECTION

Step 1. Bisect the abdominal wall (down the midline of the body). When you reach the diaphragm, stop and cut transversely to the side. This will allow the muscle tissue to flap open.

Sub-step 1. You may also have to make two transverse cuts in the lower abdominal region. In a male cat these transverse cuts must be at the level of the lower intestine, not at the hip joint. This is to avoid cutting the ductus deferens entering the abdominal cavity through the inguinal canal.

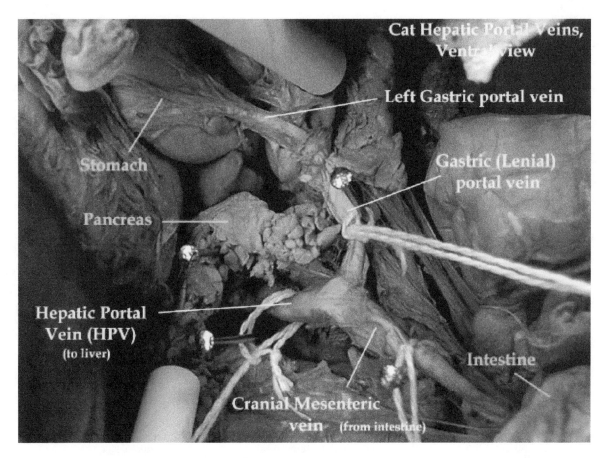

Figure 8.12 Portal veins.

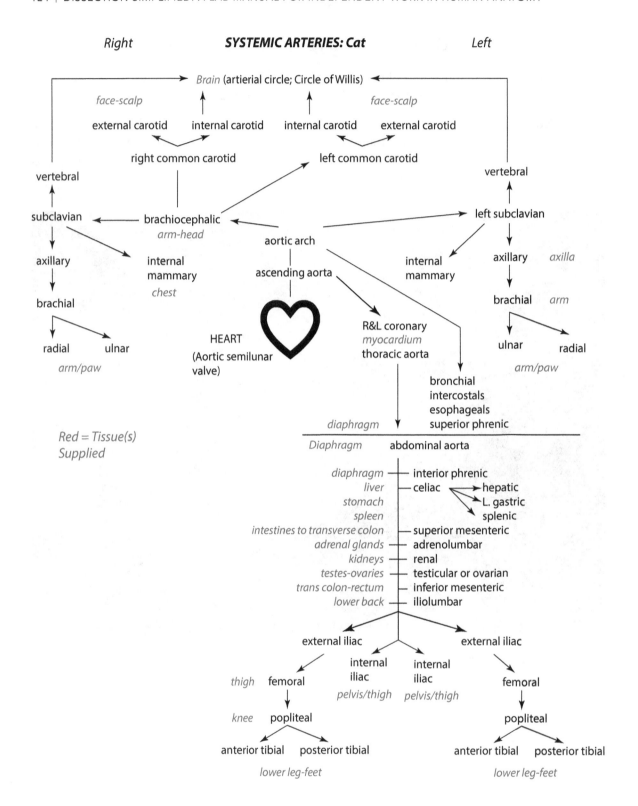

SYSTEMIC ARTERIES: Cat

Use the artery and vein diagram to label the thoracic/neck illustration.

SYSTEMIC VEINS: Cat

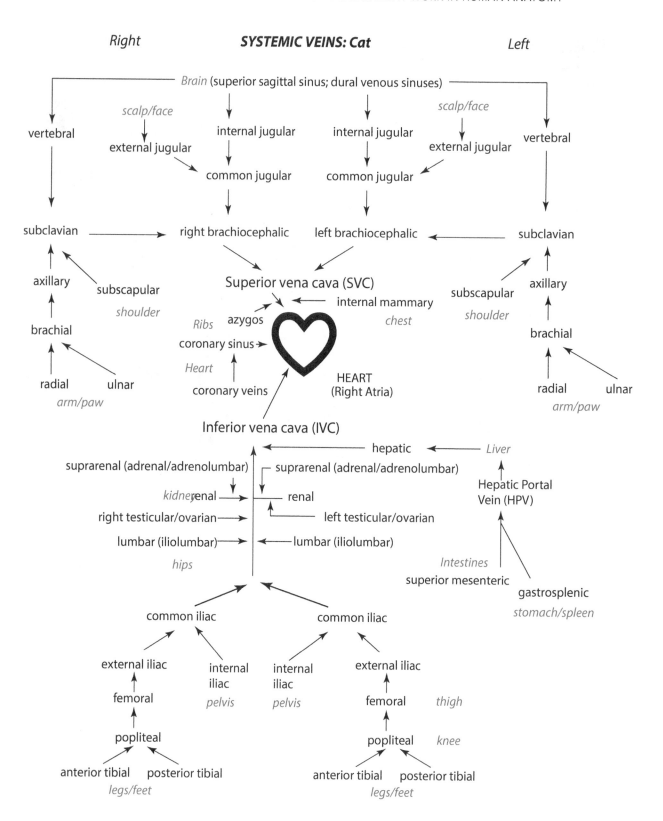

Step 2. To view the entire reproductive system, the innominate bone must be bisected along the pubic symphysis. This will require the removal of the overlying pelvic and thigh muscles. Be sure if you have a male cat to not cut the spermatic cords. Remember these are located in the pubic area.

Sub-step 2A. Find the white connective tissue line down the midline of the pubic area (underneath is the pubic symphysis.

Sub-step 2B. Insert a probe or forceps under the innominate bone to protect the tissue underneath when you cut the bone.

Sub-step 2C. Use the cutting shears to cut the innominate bone down the middle (along the pubic symphysis). The first cut will be difficult because you will be cutting the upper joining of the innominate bone. Continue cutting until you no longer hear the bone crunching.

Sub-step 2D. Once you have cut the upper 1/2 of the innominate, then you will have to cut the lower 1/2 half (the joint between the two ischium bones). Once again insert the shears to cut along the pubic symphysis and cut until the bone stops crunching.

Sub-step 2E. Now that you have the bone cut in the pubic region, it will not yet completely open because of muscle and connective tissue. To completely separate, cut along the left margin of the innominate bone. (The cat's left, not yours.) (Keep a probe or forceps under the scissors to protect the tissues underneath.)

Sub-step 2F. Note as you cut the left margin there are a few structures that can be cut: For example in male cats the **urethra** runs along underneath this area. At the end of the innominate bone in the ischium region it is very easy to cut the **penis, bulbourethral**, and the spermatic cord.

For a female cat the **vagina** and urethra run underneath this area. At the ischium region is the **urogenital sinus**, which is not as easily damaged as the penis in a male cat.

At this point of the dissection, you are ready to begin finding structures. We will start up in the facial region and work our way down.

DIGESTIVE GLANDS

I. Parotid gland Located up in the cheek region under the ear line.

II. Submandibular gland This gland is about 1/2 the size of the parotid gland and located underneath the vein running obliquely across the face (the posterior facial vein).

III. Sublingual gland This is a small triangular-shaped gland located just in front (towards the mouth) of the parotid gland. The parotid duct (a white line across the masseter) leads away from the sublingual gland.

DIGESTIVE STRUCTURES

I. Tongue You can see this structure when you cut the mental symphysis.

II. Esophagus We will see when we clear away the muscles from the trachea.

III. Stomach Located in the upper left quadrant of the peritoneal cavity. The stomach mixes food and continues chemical digestion. It contains parietal cells, which produce HCL. Chief cells (zymogenic cells) produce enzymes (pepsinogen in low pH breaks up proteins). Enteroendocrine cells secrete gastrin, which increases HCL production of parietal cells (under autonomic control).

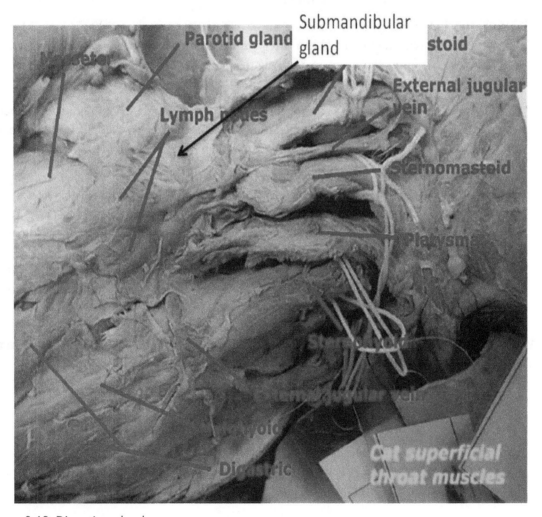

Figure 8.13 Digestive glands.

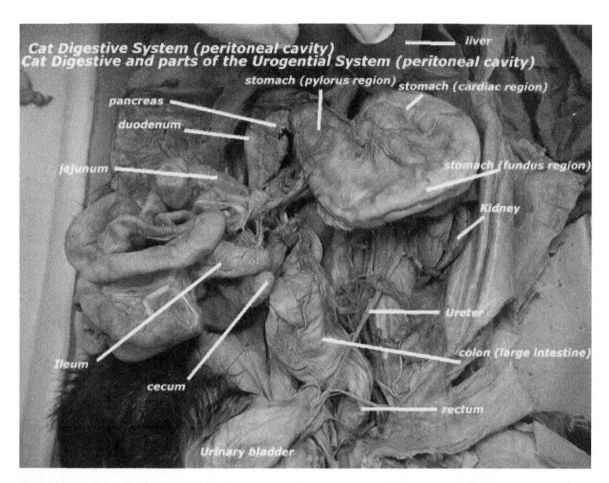

Figure 8.14 Digestive system structures and organs.

Three regions

> **Cardiac region**—top portion up where the esophagus empties into the stomach
> **Fundus**—middle outer curve; bag-shaped area
> **Pyloric**—constricted region just before the small intestine

The **greater curvature** is the outer edge of the stomach, which has the **greater omentum** attached.
The **lesser curvature** is the inner edge of the stomach and has the **lesser omentum** attached.

IV. Small Intestine
Three regions

1. The **duodenum** is the first section of the small intestine after the stomach. It is the straight portion with the pancreas nestled in the curved area. This is the major site for absorption and digestion (receives digestive enzymes via the common bile duct and pancreatic duct).
2. The **jejunum** is the second section of the small intestine in mammals. Further chemical digestion occurs in this section. This section can be seen when the small intestine begins to coil. At the first point when the straight duodenum ends the coiled jejunum begins.

3. The third and final section is the **ileum**. This section is rich with blood vessels to absorb nutrients released from the digested food. Best identified as the section of the small intestine located immediately before the ileocecal valve (located at the cecum).

V. Large Intestine (Colon) The three sections of **ascending, transverse, and descending** are difficult to distinguish in a cat. You will be required to know the colon and the rectum. The **rectum** is the lower portion of the large intestine in the region of the pelvic bone. The anus should not be difficult to find. The **colon** functions to reabsorb water and consolidate waste.

VI. Liver Part of both the digestive and endocrine system. **This organ is large and multi-lobed, located at the cranial portion of the peritoneal cavity.** The liver produces bile, which functions in the duodenum to neutralize acidity to change the pH levels to favorable conditions for digestive enzymes. The enzymes in bile are derived from hemoglobin molecules left over after red blood cells are broken down by the spleen. The liver salvages the iron from the hemoglobin. Another important function of the liver is to detoxify and phagocytize cells in the blood received from the portal system, which drains blood directly from the digestive organs.

The hepatic ducts are difficult to see and you do not have to dissect them out. But you do have to know the flow of bile from its origins in the liver to its final destination in the duodenum.

VII. Pancreas This organ has two parts. The first part can be seen in the curvature of the duodenum. The second part is more loosely attached to the mesentery under the greater curvature of the stomach. The organ has both exocrine and endocrine functions. The exocrine function is the synthesis of digestive enzymes such as pancreatic lipase, pancreatic amylase, trypsin, and chymotrypsin (endopeptidases). The endocrine function is that the pancreas secretes insulin from beta cells (found in mammals). Insulin lowers blood sugar levels by increasing the permeability of the plasma membrane of muscles and adipose tissue. Insulin also activates enzymes that convert glucose to glycogen in the liver and muscle tissue. The alpha cells of the pancreas secrete the hormone glucagon, which increases blood sugar levels by converting glycogen to glucose.

VIII. Abdominal membranes are thin and clear in appearance. These are not easy structures to pin; therefore, you will find them in written questions on the final exam. For example, the lining of the abdominal cavity is termed the _____   _____.

You would be expected to fill in the blanks with **parietal peritoneum**. So the **parietal peritoneum** is the thin lining of the abdominal wall. The visceral peritoneum is the thin membrane that covers the outer surface of organs within the peritoneal cavity. The **gastrosplenic ligament** is the membrane extending from the ventral part of the **greater omentum** and connects the stomach to the spleen. The **mesocolon** is the membrane that suspends the colon from the dorsal mesentery. The **mesentery proper** includes all the membranes interconnecting the organs of the peritoneal cavity. The best place to see this mesentery is to pull out the small intestine. Notice the thin membrane tissue with latex imbedded in it.

IX. Urogenital organs. The **kidney** should be easy to find unless the blue latex covers it over. To free the kidney, scrape the back of the peritoneal cavity with the forceps. This will loosen the membranes and connective tissue. Once the kidney is free, you should be able to see a white cord leading down the lower peritoneal area. This is the **ureter**, which drains urine into the **urinary bladder**. Be sure to bisect the kidney to see the **medulla, cortex, and renal pelvis**. The **renal papilla and pyramids** are not usually easily seen because the cut through the kidney is not completely clean.

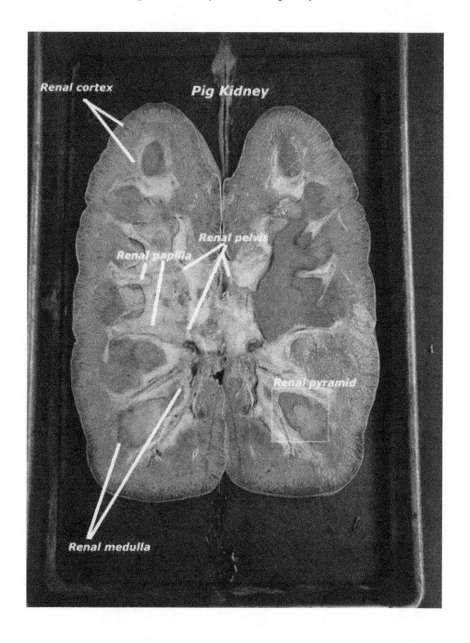

WORKSHEET IX

Blood Cells, Vessels, Heart Circulation, and Kidney Anatomy

SECTION I—BLOOD CELLS, ELEMENTS, AND VESSELS

Look at **Plates** in the lab (I–IV): these plates have the cells labeled and functions described. Once you have taken notes or looked through the known plates, use this work sheet to test your ability to identify and describe blood cells.

Use this list of blood cells and elements to fill in the blanks for questions from **Plates V–X**.

Blood elements
  Red blood cell [erythrocyte]
  Platelet
White Blood cells [leukocyte]
  Neutrophil
  Basophil
  Eosinophil
  Monocyte
  Lymphocyte

Vessels
  Artery
  Vein
  Lymphatic

Questions for Plate V:

1. Identify whether **vessel B** is an artery or vein. _____

   _____

2. Identify whether **vessel C** is an artery or vein. _____

   _____

133

3. What are **structures A**? _____.
Questions for Plate VI:

4. Identify the **unknown A** blood cell. _____

5. Identify the **unknown B** blood cell. _____

Questions for Plate VII:

6. Identify the **unknown C** blood cell. _____

7. Identify the **unknown D** blood cell. _____

8. Identify the **unknown E** blood cell. _____

Questions for Plate VIII:

9. Identify the **unknown F** blood cell. _____

10. Identify the **unknown G** blood cell. _____

11. Identify the **unknown H** blood cell. _____

12. Identify the **unknown J** blood cell. _____

Questions for Plate IX:

13. Identify the **unknown vessel in figure 1**. _____

14. Identify the **unknown vessel in figure 2**. _____

15. Identify the **unknown vessel in figure 3**. _____

16. Identify the **unknown vessel in figure 4**. _____

Questions for Plate X:

17. Plate X illustrates macrophage cells after they have engulfed carbon particles that were loose within lung tissue. Which type of white blood cell serves this macrophage function?

18. Which white blood cell forms mast cells after exiting the circulatory system to enter connective tissue? These cells also release heparin, histamine, and serotonin.

_____

[an association with this behavior and the white blood cell: histamine is released during allergic reactions, which is a **basic** response to irritating particles in the air]

19. Which white blood cell secretes chemicals and destroys invading bacteria?

_____

[an association with this behavior and the white blood cell: these cells **neutralize** bacteria.

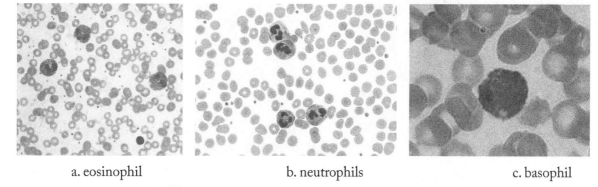

a. eosinophil   b. neutrophils   c. basophil

Figure 9.1 Types of white blood cells.

9.1a: Copyright © Ed Uthman (CC by 3.0) at http://commons.wikimedia.org/wiki/File:Eosinophils_in_peripheral_blood.jpg.
9.1b: Copyright © Dr. Graham Beards (CC BY-SA 3.0) at http://commons.wikimedia.org/wiki/File:Neutrophils.jpg.
9.1c: Copyright © Reytan (CC BY-SA 3.0) at http://commons.wikimedia.org/wiki/File:Basophil.jpg.

20. Which of the white blood cells illustrated in Figure 9.1 will increase in number if the body is infected with a parasitic worm?

These same cells also phagocytize antibody-antigen complexes.

21. Which type of lymphocyte differentiates into a plasma cell that can produce antibodies?

_____

22. The thymus gland produces a type of lymphocyte that directly attacks foreign cells including cancer cells, bacterial cells, fungal cells, viruses, and cells from transplanted organs. The designation for this lymphocyte is _____.

SECTION II—HEART STRUCTURES AND CIRCULATION

View the demo heart and identify the labeled structures.

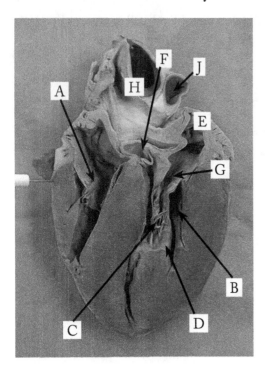

Figure 9.2 Internal view of the Posterior heart section.

A [chamber]: _____

B [chamber]: _____

C: _____

D: _____

E: _____

F: _____

G: _____

1. The tricuspid valve is between which two chambers?

_____

2. The coronary sinus empties into which chamber?

_____

3. This is a cut anterior view of a pig heart. Identify the cut openings labeled H and J.

_____

_____

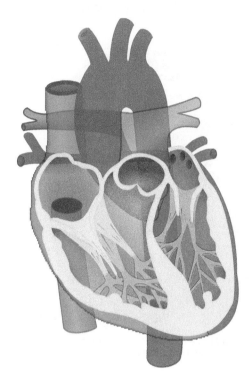

Figure 9.3 Heart blood flow diagram.

Copyright © ZooFari (CC BY-SA 3.0) at http://commons.
wikimedia.org/wiki/File:Heart_diagram_blood_flow_en.svg.

Use the diagram and fill-in-the blanks for the direction of blood flow through the heart.

1. Vessels that drain deoxygenated blood into the heart [from the body and heart]

_____

_____

_____

2. [chamber]: _____

3. [valve]: _____

4. [chamber]: _____

5. [valve]: _____

6. [vessel(s)]: _____

7. [organ]: _____

8. [vessel(s)]: _____

9. [chamber]: _____

10. [valve]: _____

11. [chamber]: _____

12. [valve]: _____

13. [vessel]: _____

Heart structure questions:

14. The term for the fetal remnant connection between the pulmonary trunk and aortic arch.

_____

15. The term for the fetal remnant of the foramen ovale between right atria and left atria.

_____

SECTION 3—MACRO/MICRO ANATOMY OF THE KIDNEY

Use the diagram and fill in the blanks for the structures labeled.

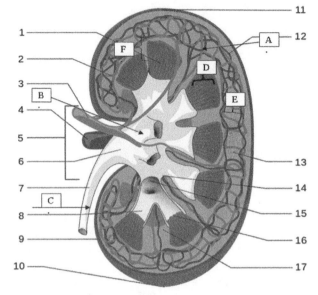

3. _____

B. _____

C. _____

Copyright © Piotr Michal Jaworski (CC BY-SA 3.0) at http://commons.wikime-dia.org/wiki/File:KidneyStructures_PioM.svg.

Nephron Structure: use the diagram and fill in the blanks for the structures labeled.

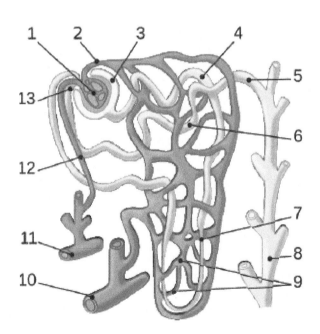

1. _____

2. _____

3. _____

4. _____

6. _____

7. _____

Burton Radons, "Kidney Nephron," https://commons.wikimedia.org/wiki/File:Nephron_illustration.svg. Copyright in the Public Domain.

SECTION 4—REVIEW IDENTIFICATION BLOOD AND VESSELS

After completing the unknown blood cells and vessels identify the following unknowns.

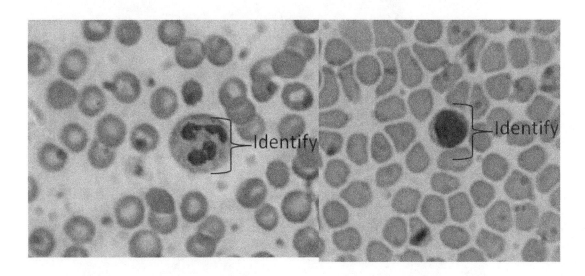

Blood component A   Blood component B

Identify the blood cell component A. _____
Identify the blood cell component B. _____

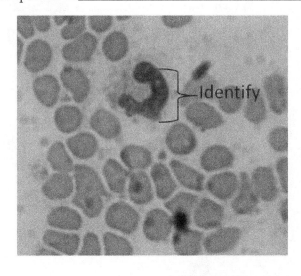

Blood component C

Identify the blood cell component C. _____

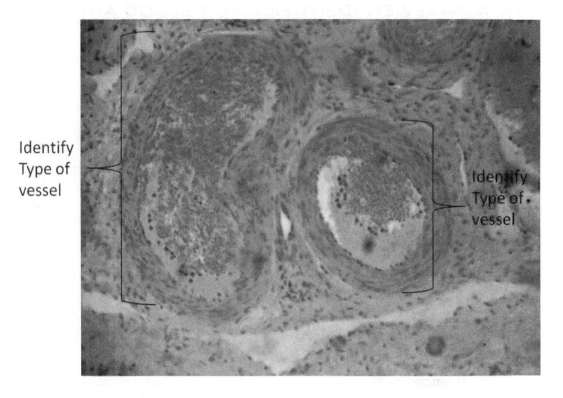

Identify Type of vessel

Identify Type of vessel

Identify the type of vessel: Which one is an artery? _____

Which one is a vein?_____

Identify the type of vessel illustrated_____.

Figure 9.4 Lateral throat view.

SECTION 5—LABEL VESSELS AND ORGANS

Identify organs and vessels in the photo by labeling the following.

Vessels in view: anterior facial, posterior facial, transverse jugular, external jugular, axillary vein, Subclavian vein, brachiocephalic vein, superior vena cava, left and right common carotids, internal mammary artery, final portion of the left Subclavian artery

Organs in view: Submandibular gland, parotid gland, esophagus.

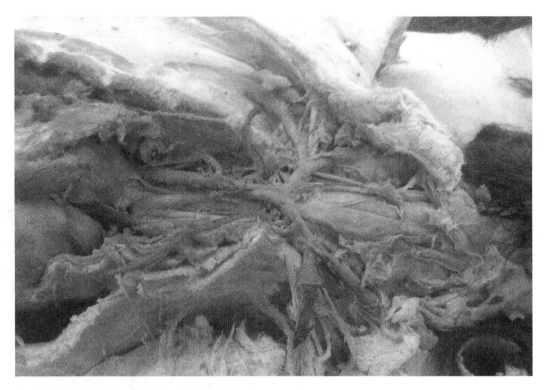

Figure 9.5 Anterior thoracic view.

Identify organs and vessels in the photo by labeling the following.

Vessels in the view to identify include: superior vena cava, internal mammary vein, brachiocephalic vein, Subclavian vein, axillary vein, external jugular vein. Brachiocephalic artery, left Subclavian artery, internal mammary artery, axillary artery, right and left common carotid.

Organ in view to label include: Lungs, trachea, esophagus, tracheal rings (structure not organ), Heart (with pericardial cavity intact). The thymus was removed to show the superior vena cava.

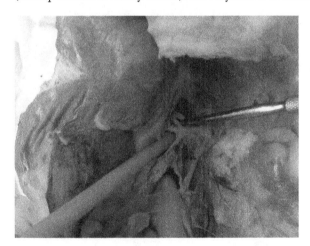

Identify the following in Figure 9.6:

Celiac trunk, superior mesenteric artery, hepatic artery, left gastric artery, splenic artery, renal vein, and inferior vena cava Organs: spleen pancreas, kidney, stomach

Figure 9.6 anterior view of abdominal vessels

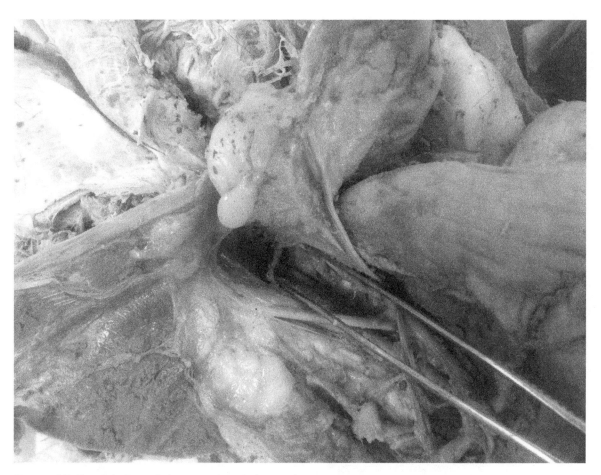

Figure 9.7 Caudal view of the anterior abdominal.

Identify organs and vessels in the photo by labeling the following.

Vessels to label in view include: internal iliac vein, external iliac vein, common iliac vein, inferior vena cava, and external iliac artery.

Organs in view to label include: urinary bladder, ureter, and uterine horns.

WORKSHEET X

Nervous System, Eye, and Ear

Locate the following structures on the sheep brains and human brain models

Embryonic brain section	CNS structure	
Telencephalon	cerebral hemispheres	Differentiate the: frontal lobe, parietal lobe, occipital lobe, and temporal lobe
	cerebral cortex	
	corpus callosum	
	olfactory bulb	
Diencephalon	**thalamus**	
	hypothalamus	
	pineal gland	
	pituitary gland	
Mesencephalon	**cerebral peduncles**	
	corpora quadrigemina	
Metencephalon	**cerebellum**	
	pons	
Myelencephalon	**medulla oblongata**	

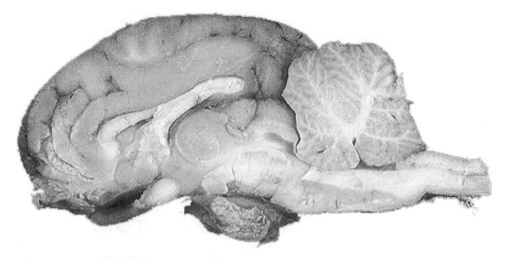

Figure 10.1 Sagittal section of sheep brain to label. Label all structures from list, which include: Corpus callosum, Cerebral hemisphere, thalamus, pineal gland, hypothalamus, pituitary, corpora quadrigemina (with superior colliculi), cerebellum, cerebral peduncle (more in a ventral view), Pons (also ventral view), and medulla oblongata (also ventral view).

Brain Questions:

1. Which structures comprise the brain stem?
2. The brain is surrounded by supporting layers of membranes referred to as the meninges. What are the three layers?

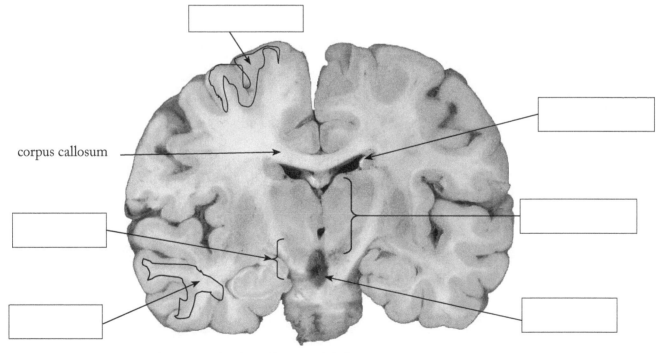

corpus callosum

Figure 10.2 Coronal section of a sheep brain to label.

3. Which layer are you likely to see on the sheep brain specimens? [Hint: try to separate the gyri on the cerebrum.]

Brain Structure Functions:

4. What is the general function of the hypothalamus?

5. Which of the following would not be stimulated by neurons originating in the hypothalamus?

    a. contraction of smooth muscle cells in the intestinal tract
    b. the sensation of thirst
    c. motor neurons controlling the eye muscles
    d. autonomic neurons to stimulate a shiver response in muscle fibers

6. What is the general function of the thalamus?

7. The thalamus is responsible for _____?

    a. the somatic control of arm muscles
    b. regulation of a student's alertness level during lecture
    c. the somatic reflex response
    d. stimulating an aggressive response to auditory stimuli

8. What is the general function of the pons?

9. Which of the following would not involve neurons originating in the pons?

    a. regulation of breathing actions
    b. sensing a loss of balance
    c. control of eye movements
    d. regulation of cardiac rhythm

10. Which brain structure contains the neural cell bodies of the vagus nerve?

Brain Ventricles and cerebral spinal fluid: (lateral ventricles and third ventricle can be found on the sheep brain)

Ventricles of the brain
   Lateral ventricles
   Ventricle III
   Cerebral Aqueduct (of Sylvius)
   Ventricle IV
   Choroid plexus

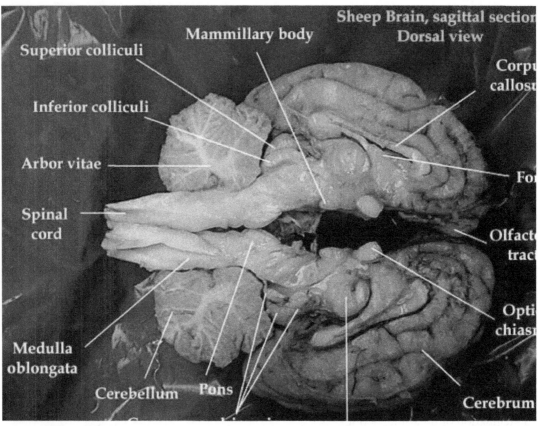

Figure 10.3 Labeled sagittal sheep brain.

11. Which structure(s) is responsible for producing cerebral spinal fluid?

12. How does cerebral spinal fluid differ from blood?

13. How is the cerebral spinal fluid absorbed back into the circulatory system?

14. How is the cerebral spinal fluid circulated around the brain and spinal cord?

BLOOD VESSELS OF THE BRAIN

<u>Identify the following:</u>

Arterial circle (of Willis)* locate on the sheep brain
basilar artery* locate on the sheep brain
vertebral arteries
common carotid arteries (pass through the carotid foramina)

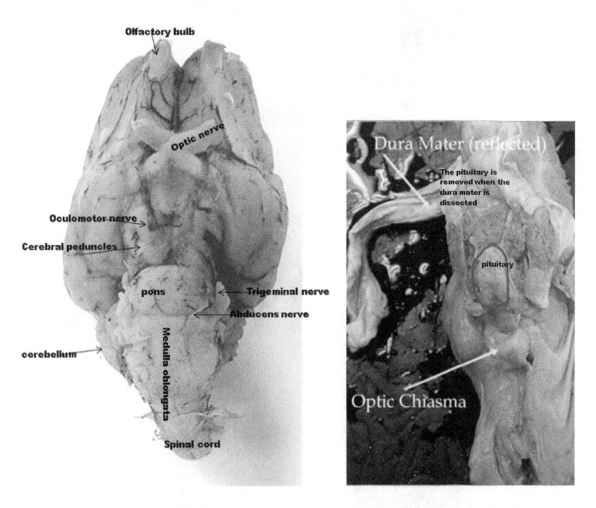

Figure 10.4 Labeled ventral sheep brain with pituitary and without the dura mater and pituitary.

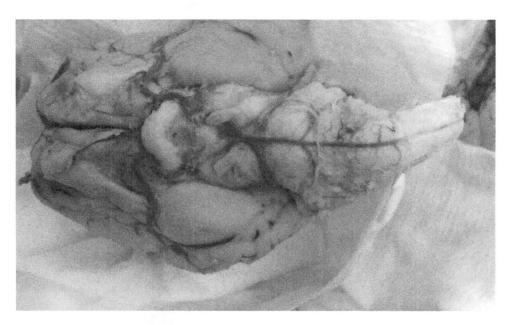

Figure 10.5 Ventral view of sheep brain. Label all structures, cranial nerves and vessels.

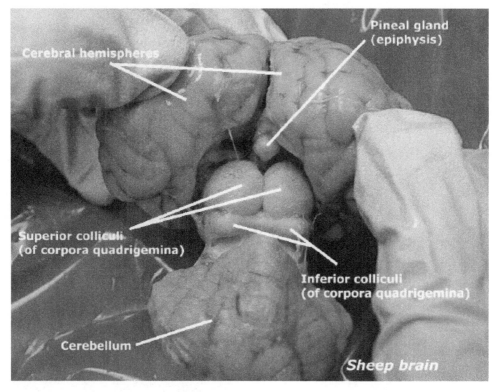

Figure 10.6 Labeled posterior view of the sheep brain.

Peripheral Nervous system: Cranial nerves

Locate the following individual nerve stumps on the sheep brain:
olfactory, optic, oculomotor, trigeminal, and abducens

The following are difficult to distinguish individually on the cat brain, so don't try: trochlear, glossopharyngeal, vagus, spinal accessory, hypoglossal

	Nerve	**Foramen**
I.	Olfactory	holes in cribriform plate of ethmoid
II.	Optic	optic foramen in sphenoid
III.	Oculomotor	superior orbital fissure
IV.	Trochlear	superior orbital fissure
V.	Trigeminal	V_1—foramen ovale
		V_2—foramen rotundum
		V_3—superior orbital fissure
VI.	Abducens	superior orbital fissure
VII.	Facial	internal auditory meatus, stylomastoid foramen
VIII.	Auditory (vestibulocochlear)	internal auditory meatus
IX.	Glossopharyngeal	jugular foramen
X.	Vagus	jugular foramen
XI.	Spinal accessory	jugular foramen
XII.	Hypoglossal	hypoglossal foramen

Locate the following cranial nerve outside the braincase, on the cats

Vagus

Locate the following cranial nerve outside the braincase, the available models

Optic, Olfactory, Trigeminal, Abducens, Facial, Auditory, Vagus, spinal accessory, and hypoglossal.

Nervous system Assignment:

15. What are the three initial embryonic sections of the central nervous system?

16. Two of the initial three embryonic sections differentiate resulting in five sections to the brain. The thalamus as part of the diencephalon develops from which initial embryonic brain section [note there are three embryonic brain sections from which five sections diverge].

17. The central nervous system develops from a linear arrangement of cells that migrate inward to form the neural tube. List the five development segments in order anterior to posterior.
What are the primary functions of the brain structures of the diencephalon?

Number	S/M/B	Name	Location of cell bodies	Mnemonic to remember	Mnemonic to remember
CN-I	S	Olfactory	Telencephalon	On	Some
CN-II	S	Optic	_____	Old	Say
CN-III	M	Oculomotor	mesencephalon	Olympus	Marry
CN-IV	M	Trochlear	_____	Towering	Money
CN-V	B	Trigeminal	Pons _____	Tops	but
CN-VI	M	Abducent	_____	A	my
CN-VII	B	Facial	_____	Fin	brother
CN-VIII	S	Vestibulocochlear	_____	And	says
CN-IX	B	Glossopharyngeal	_____	German	bad
CN-X	B	Vagus	_____	View	boys
CN-XI	M	Spinal Accessory	_____	Some	marry
CN-XII	M	Hypoglossal	_____	Hops	money
			Brain stem/medulla		

18. Fill in the locations of the cells bodies for the missing cranial nerves in the above table. The table also includes two suggested mnemonics for learning the cranial nerves by Roman numeral, name, and neuron function. You will be required to identify a cranial nerve by either Roman numeral or by name and therefore need to know them by both.

For example: Identify a cranial nerve, by name, that exits the foramen ovale. _____

Identify which of the following would have only motor neurons within the nerve. A. II, III, and IV B. III, XI, and XII C. IX, II, and X (circle the correct choice).

There are associations with many of the cranial nerves that will help remember the Roman Numeral with the name.

A. You smell something before you see it. When you look to see the source of the smell you move your eyes. These facts will help order the three "O"s. I-Olfactory II Optic III Oculomotor
B. Trigeminal - Tri = 3 gemini = the twins or 2   3 + 2 = 5
C. 6-pack for abs - The sixth cranial nerve is the abducens.
D. I associate the face having flesh removed with the movie *Seven*—that association helps me remember the seventh cranial nerve is the Facial.

These are a few I use to remember the cranial nerves and their associated Roman numeral. You should find your own as they will be easier for you to remember. Associating new information with previous memories will increase the ability to recall the newer information.

19. Cranial nerve VI [abducens] is located within which of the five developmental regions of the brain?

20. Cranial nerves XI and XII are located within which of the five developmental regions of the brain?

21. Outline or shade the sections in the illustration all visible structures of the diencephalon.

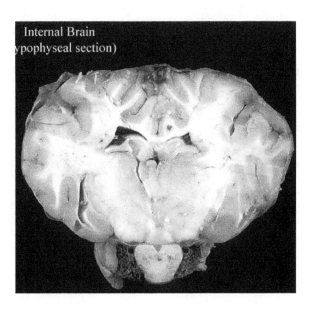

Internal Brain
(hypophyseal section)

22. What is the primary function of the myelencephalon?

Nervous system topic: Cranial nerves and the foramen they exit

23. Which foramen can be seen in an inferior view of the ventral skull?

24. If you insert a wooden dowel through the optic foramen [from the internal superior skull view] the dowel will exit through into which structure? _____.

25. Which foramen can be seen only in a superior internal skull view?

26. Which nerves exit through the jugular foramen? Write out their roman numerals and name of the nerve

27. The olfactory nerve I , exits which foramina?_____ These foramina are located in what structure? _____.

Nervous system topic: Cranial nerves and the structures or organs they innervate [functions: sensory, motor or both sensory and motor]

28. Identify the cranial nerves with neurons that only transmit sensory information.

29. Identify the cranial nerves with neurons that only transmit motor information.

30. What anatomical structures does the glossopharyngeal nerve innervate?

31. What anatomical structures does the Vagus nerve innervate?

32. If the spinal accessory nerve were tagged on a cat specimen which muscle would be visible at its terminal innervation?

<u>Peripheral Nervous system</u>

On the cats, human vertebral column, and spinal nerve model, locate the:

Brachial Plexus
Identify the following nerves:
phrenic, radial, median, ulnar   (identify the brachial nerves on the cat)

Lumbosacral plexus
Identify the following nerves:
femoral, obturator, sciatic

Study the model of a neuron and identify the following:

Schwann cells (myelin sheath), nerve cell body, axon, dendrites,
axon terminals (synaptic bulbs), nucleus, cytoplasm

<u>Plate V</u>

33. Identify structure A _____

34. Identify cell B _____

35. Identify structure C _____

36. Identify component D _____

Study the model of the cervical spinal nerve:

Locate the parts of the vertebra including the articular facets, and body and spinous process; locate the spinal cord (white and gray matter), dura mater, pia mater, spinal nerve (posterior root and posterior root ganglion-sensory, anterior root-motor, dorsal ramus, ventral ramus), sympathetic trunk (chain and ganglion), communicating rami, vertebral artery and vein

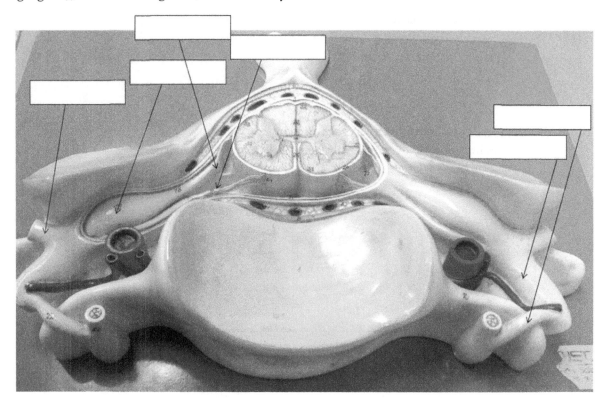

Figure 10.7 Labeled spinal cord model.

SPECIAL SENSES: EYE AND EAR

Orbit and Its Contents

On the cat eye and plastic eye model, locate the following:

Extrinsic eye muscles (innervated by cranial nerves III, IV, VI)
Optic Nerve

Identify using the sheep eye the following
        sclera, cornea, lens, iris, pupil, conjunctiva, ciliary body,
        anterior chamber, posterior chamber
        (aqueous humor), (vitreous humor), retina, pigmented layer

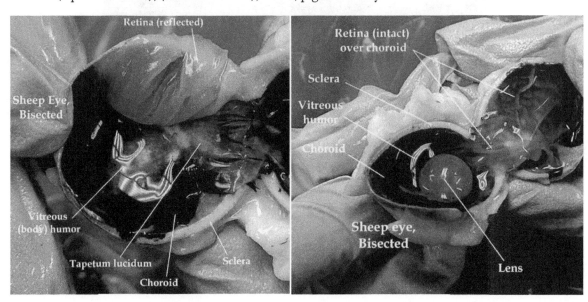

Figure 10.8 Internal views of the sheep eye.

Use the textbook to identify the following on the human ear model

External auditory meatus
Internal auditory meatus
Tympanic membrane (ear drum)
Ossicles: incus, malleus, stapes (in block of plastic) and demonstration
Eustachian tube (auditory tube)
Semicircular canals
Vestibule: utricle and saccule
Cochlea

Oval window
Round window
Auditory nerve (equals vestibulocochlear nerve)

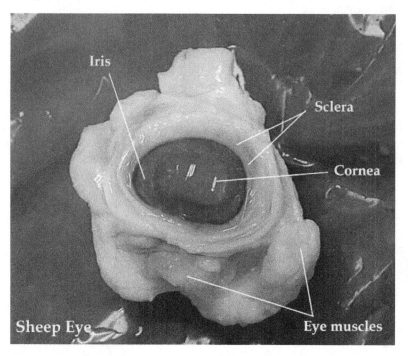

Figure 10.9 External sheep eye.

WORKSHEET XI

Reproductive System and Embryo Development

REPRODUCTIVE SYSTEM

<u>Identify the following on the cat and female pelvic model</u>

Uterus, Ovary, Vagina, Urogenital vestibule [sinus], horns (cornua: found in cat, not human), uterine tube (fallopian tubes)

<u>Identify on the female pelvic model</u>

broad ligament, ovarian ligament, round ligament, cervix, clitoris.

<u>Identify in a cross section slide of an ovary:</u>

<u>Differentiate between the following follicles:</u>

   —primodial
   —primary
   —secondary
   —graafian

<u>Identify the following within a follicle:</u>

Granulosa cells
Ovvum
zona pellucida
theca

The Primordial Follicle

The primordial follicle can be identified by its single layer of follicular cells (red arrow). To the right are two atretic follicles (blue arrows).

Atretic follicles are ones whose development has been curtailed. Usually only one follicle will complete the entire process to maturity and ovulation.

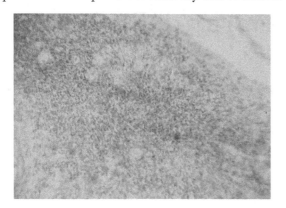

Primary Follicle

FSH from the adenohypophysis stimulates follicular development. At least two layers of follicular cells identify the primary follicle.

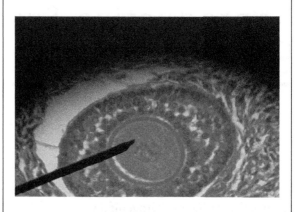

Adapted from: "Primary Follicle," https://commons.wikimedia.org/wiki/File:Primary_follicle-4.JPG. Copyright in the Public Domain.

Secondary Follicle

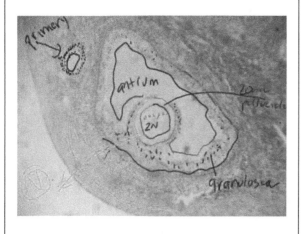

Graafian Follicle

The Graafian follicle is identified by the large antrum (A) and the corona radiata (arrow) that surrounds the actual oocyte and projects into the antrum.

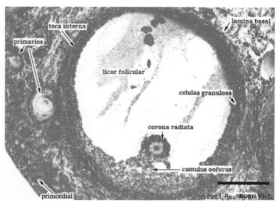

Copyright © Tufts Open Courseware (CC by 3.0) at https://commons.wikimedia.org/wiki/File:Foliculo_de_Graaf.png.

SECTION 11 A—THE FEMALE REPRODUCTIVE SYSTEM

Label female reproductive histology structures and identify follicle type.

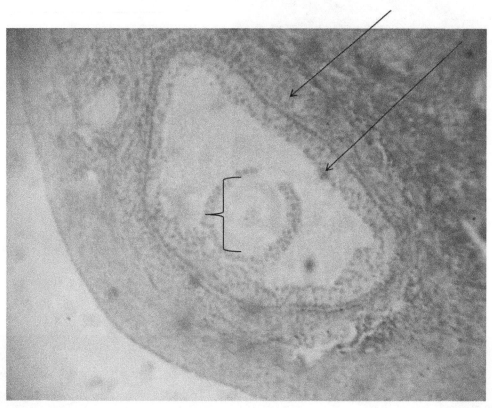

Figure 11A Label follicle cells.

Follicle type?_____

How many chromosomes are within the oocyte?_____

Identify Follicle

A _____

B _____

C _____

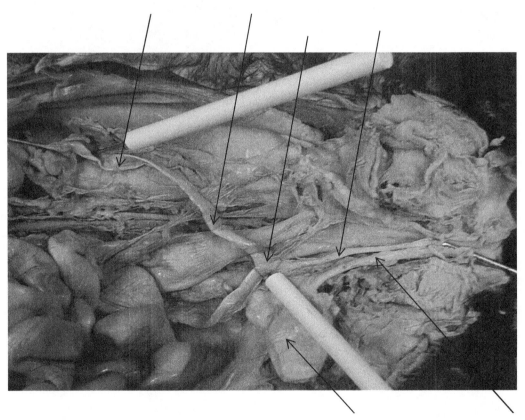

Figure 11.1 Female reproductive structures to label. Identify female organs and structures in view. Complete labeling for credit. Organs in view to identify are: ovary, fallopian tube (very small), uterine horns, body of the uterus, vagina, urinary bladder, and urethra.

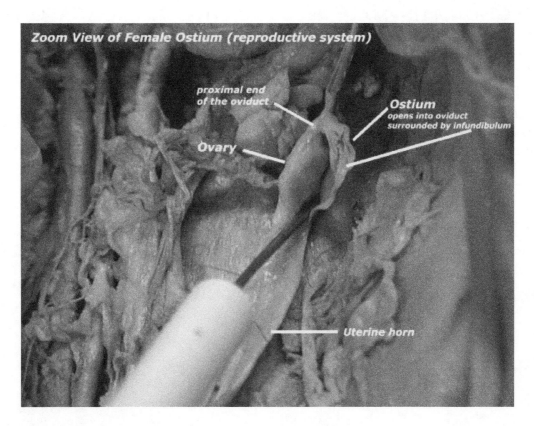

Figure 11.2 Female cat ovary with ostium.

The Female Reproductive System

The ovary is a small bean shaped organ which may be difficult to find at first because of the connective tissue around it. The best way to find the ovary is to follow the uterine horn up. These are the thin strips of beige cord tissue on either side the peritoneal cavity.

The **osteum and fimbrae** are difficult to see so you should just know them by description. The **osteum** is the opening into the oviduct and **the fimbrae** are the small finger projections along the edge of the oviduct.

The **oviduct** in cats is not very long. It is a small thin tube that runs from the top area of the ovary to the bottom area of the ovary and connective tissue. The major part of the reproductive tract you can see is the uterine horn. The **uterine horn** is where a fertilized egg will develop.

The two **uterine horns** lead to the small **body of the uterus.** It is the small triangular tissue at the base of the two uterine horns. At the base of the uterus is the **cervix.** The **cervix is** an internal structure and you do not need to dissect out the reproductive tract. Just know the location of the cervix.

The **Vagina** is the continuation of the reproductive tract after the uterus. The vagina empties into the **urogenital sinus.** Note this is also where the urethra empties into.

SECTION 11B—MALE REPRODUCTIVE MACROANATOMY AND HISTOLOGY

<u>Identify in a cross section slide of a testis:</u>

Semniferous tubules
Interstitial cells
Primary spermatocytes
Secondary spermatocytes

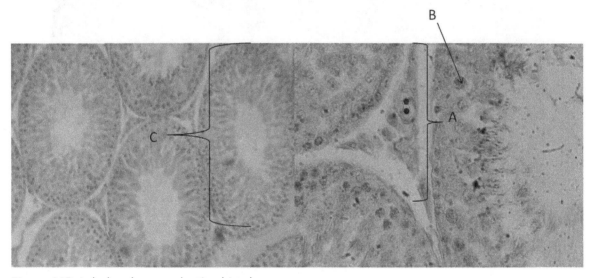

Figure 11B Label male reproductive histology.

Identify Cells A _____

Identify Cells B _____

Are cells B diploid or haploid? _____

Identify structure C _____

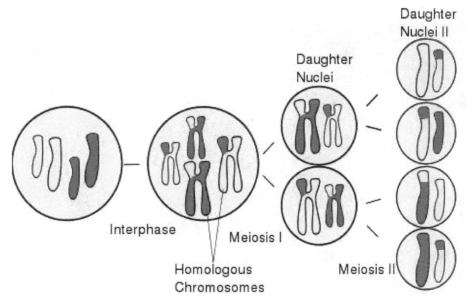

Ultra two, "Meiosis diagram," https://commons.wikimedia.org/wiki/Category:Meiosis#/media/File:MajorEventsInMeiosis.svg. Copyright in the Public Domain.

Cells B are at which step in Meiosis?_____

Which cells would be at Meiosis II?_____

Identify the following on the cat and the male pelvic model

Penis, scrotum, testes, vas deferens, epididymis
Inguinal canal, Prostate gland, Urethra

Identify on male pelvic model

Cavernous bodies (corpora cavernosa)
Spongy bodies (corpus spongiosum)
Seminal vesicle

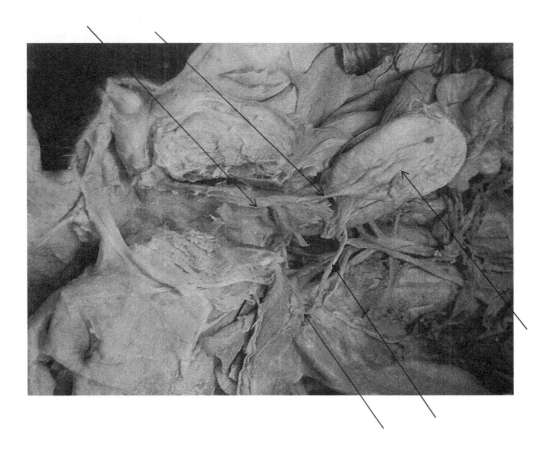

Figure 11.3 Male reproductive illustration to label Organs and structures to label include the following: Prostate, urethra, urinary bladder, vas deferns, Inguinal canal.

The Male Reproductive System

The **scrotum** is the skin covering over the **testes**. You will have to cut the skin tissue so as to split the **scrotum** open. Peal the outer scrotum covering away from the testes. (Hopefully you have not cut the spermatic cord so the testes will be attached to the body by the spermatic cord even after being removed from the scrotum.

There is a white connective tissue covering over the testes, this is the **tunica vaginalis**. Cut a slit in the tunica vaginalis and open it up to expose the testis inside.

The **epididymis** is the coiled portion of the first portion of the duct system draining the testis. You can identify the **epididymis** because it will appear as a flap curved around the testis. Once the duct appears not to be coiled (this happens as the duct leaves the area of the testis), at this point the duct is referred to as the **ductus deferens**.

The **ductus deferens** enters the peritoneal cavity by way of the inguinal canal. It then loops around and dives down the the pelvic region. At the point where the two ductus deferens meet is the location of the **prostate gland.** This is also the location of the **urethra.**

The **bulbourethral glands** are located lateral to the urethra at the point of the ischium. After the end point of the ischium the urethra enters **the penis.** You will have to cut the overlying skin to see this distinction.

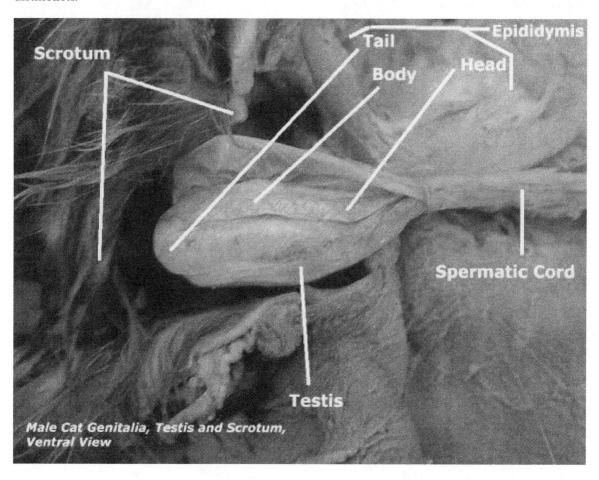

Figure 11.4 Male reproductive organs.

EMBRYOLOGY

There are a set of embryo plates available during the lab (I–VIII); use these plates to answer the following question set.

Early stages: Following fertilization of the egg, **cleavage** gives rise to 2 cells, 4 cells, 8 cells, 16 cells, and 32 cells. At this stage the embryo consists of superficially similar cells. Then subdivision occurs into the inner cell mass and the trophoblast, giving rise to cells destined to form the embryo and the extra-embryonic tissues, respectively. It takes about a week for these cleavages to occur and for the embryo to move down the Fallopian tube into the uterus, where it will implant in the wall of the uterus. At about the time of implantation, the zygote undergoes embryogenesis and the inner cell mass is subdivided into two tissue layers and the embryo forms a disc.

Use Plate I to answer the following questions:

1. Identify the cell stage when the fertilized zygote implants in the wall of the uterus. _____

2. Refer to Figure 2; the area outlined and labeled E contains _____.
   A. the trophoblast cells   B. the inner cell mass
   C. cleavage cells   D. a Morula cell mass

3. Which two cell layers develop in the inner cell mass after implantation?

Use Plate II to answer the following questions:

4. Which of the labeled structures (A–E) will contribute cells to the developing chorion?

5. The chorion serves which of the following functions in a developing fetus?
   A. embryonic disc cells   B. embryonic area of gas and waste exchange the placenta
   C. fluid-filled cavity protecting   D. forming the yolk sac and cells contributing to
      the embryo   the primitive gut.

6. Which labeled structure is the developing amniotic cavity?

Gastrulation transforms this two-layered organization into a three-layered one. At the onset of gastrulation, the embryonic disc becomes elongated and is wider at the cranial end. The primitive streak runs from the future caudal end of the embryo in a cranial direction. In this region, cells from the top cell layer move through the streak to contribute to the mesoderm and the endodermal germ layers. Remaining cells in the top layer contribute to the ectodermal germ layer. At the region of the future mouth and anus, mesoderm cells fail to intervene between ectoderm and endoderm, which remain closely apposed and this region eventually breaks down. Thus at the end of gastrulation, the future head and tail are defined, and the dorsoventral axis is defined by the primitive streak dorsally and the lateral edges of the embryonic disc ventrally.

Use Plate III to answer the following questions:

7. Identify structure A _____

8. Identify structure B _____

9. Identify structure C _____

10. Identify area D _____

11. The chick embryo illustrated is approximately 24 hours into development; there is an equivalent human embryo stage illustrated. The human embryo appears to be approximately how many hours/ days into development?
    A. 50 days
    B. 20 days
    C. 20 hours
    D. 50 hours

12. The presence of structure C indicates which of the following process(es) have occurred during development?
    A. neurulation
    B. gastrulation
    C. placentation
    D. embryonic folding
    E. both A and B.

Neurulation begins as gastrulation is completing. The dorsal region of the ectoderm [**neural plate**] becomes thickened and sinks down in the dorsal midline, while two neural folds rise up on each side of the midline. These folds eventually meet one another and start to fuse at the future neck region. Closure of the neural folds then progresses cranially and caudally. The **neural tube** that is thus formed separates from the overlying ectoderm destined to form the epidermis. This tube shows expansions cranially, which will form the forebrain, midbrain, and hindbrain. Associated with neurulation, the mesoderm becomes separated into several components—the **notochord** dorsally and the **paraxial mesoderm** adjacent to it. In the trunk the paraxial mesoderm is divided **into somites**, from which derive all the muscles, the dorsal axial muscles, ventral body wall muscles, **and the limb muscles. Lateral to the paraxial mesoderm is the intermediate and lateral plate mesoderm.**

    Neural crest cells are formed at the junction between the neural plate and epidermis during gastrulation. These cells are migratory and will form neurons in the peripheral nervous system, as well as a large number of other important derivatives. They can be considered as the fourth germ layer. **Folding** of the embryo in the craniocaudal and the dorsoventral directions lead to enclosure of the endoderm and regionalization of the gut tube into foregut, midgut and hindgut. By the end of the first month, the embryo has a recognizable form and the major coordinates of the body (craniocaudal and dorsoventral axis) have been set up.

Use Plate IV to answer the following questions:

13. Identify structure A _____

14. Identify structure B _____

15. Identify region C _____

16. Identify region D _____

17. Identify structure E _____

18. Which of the following process(es) was not completed by the stage illustrated in Plate III but has occurred by the stage of development illustrated in Plate IV?
    A. neurulation   B. gastrulation
    C. placentation   D. A and B.

Use Plate V to answer the following questions:

19. Identify structure A _____

20. Identify structure B _____

21. Identify structure C _____

22. Identify structure(s) D _____

23. The chick embryo illustrated is approximately 56 hours into development; there is an equivalent human embryo stage illustrated. The human embryo appears to be approximately how many hours/days into development?
    A. 32 days   B. 52 days
    C. 32 hours   D. 50 hours

24. Although all vertebrates have pharyngeal arches during their embryonic development, humans do not retain the pharyngeal arch structure. The pharyngeal arches of a human embryo develop into

    _____.

25. At the fifth week, is the developing human considered an embryo or a fetus? _____

Use Plate VI to answer the following questions:

26. During which week of human development does neurulation occur? _____

27. True or False; the notochord structure will become the spinal cord. _____

28. What structures will the somites differentiate into? _____

Use Plate VII to answer the following questions:

29. Describe the difference between neurulation and embryonic folding.

_____

30. True or False; during lateral folding of the embryo the yolk sac becomes smaller and the amniotic cavity becomes larger. _____

31. Refer to the figure of vertebrate embryology, row II depicts a human embryo at about _____.

    A. 2 days of development   B. 32 days of development   C. 62 days of development.

32. What is the cause of embryonic folding? _____